Sustainable Farmland Management: Transdisciplinary Approaches

Sustainable Farmland Management: Transdisciplinary Approaches

Edited by

Robert Fish
Centre for Rural Policy Research, University of Exeter, UK

Susanne Seymour
School of Geography, University of Nottingham, UK

Charles Watkins
School of Geography, University of Nottingham, UK

Michael Steven
School of Geography, University of Nottingham, UK

www.cabi.org

CABI is a trading name of CAB International

CABI Head Office
Nosworthy Way
Wallingford
Oxfordshire OX10 8DE
UK

Tel: +44 (0)1491 832111
Fax: +44 (0)1491 833508
E-mail: cabi@cabi.org
Website: www.cabi.org

CABI North American Office
875 Massachusetts Avenue
7th Floor
Cambridge, MA 02139
USA

Tel: +1 617 395 4056
Fax: +1 617 354 6875
E-mail: cabi-nao@cabi.org

A catalogue record for this book is available from the British Library, London, UK.

Library of Congress Cataloging-in-Publication Data

Sustainable farmland management: transdisciplinary approaches / edited by Robert Fish ...[et al.].
p. cm.
Includes bibliographical references and index.
ISBN 978-1-84593-351-7 (alk. paper)
1. Sustainable agriculture. 2. Farm management. I. Fish, Robert. II. Title.

S494.5.S86S884 2008
630--dc22

2008015937

ISBN: 978 1 84593 351 7

Printed and bound in the UK by MPG Books Limited, Bodmin, from copy supplied by the editors.

Contents

List of Contributors

Konstantinos N. Baginetas: *Department of Agriculture, Lamia, Greece.*
Email: kbaginetas@yahoo.co.uk

John V. Beckett: *School of History, University of Nottingham, University Park, Nottingham, NG7 2RD, UK.*
Email: john.beckett@nottingham.ac.uk

John Boardman: *Environmental Change Institute, Oxford University Centre for the Environment, South Parks Road, Oxford, OX1 3QY, UK.*
Email: john.boardman@eci.ox.ac.uk

Henry Buller: *Department of Geography, School of Geography, Archaeology and Earth Resources, University of Exeter, The Queen's Drive, Exeter, EX4 4QJ, UK.*
Email: h.buller@exeter.ac.uk

Caroline Drummond: *LEAF, NAC, Stoneleigh, Warwickshire, CV8 2LZ, UK.*
Email: caroline.drummond@leafuk.org

Chris Firth: *Henry Doubleday Research Association (HDRA), Ryton Organic Gardens, Coventry, CV8 3LG, UK.*
Email: cfirth@hdra.org.uk

Robert Fish: *Centre for Rural Policy Research, University of Exeter, The Queen's Drive, Exeter, EX4 4QJ, UK.*
Email: R.D.Fish@exeter.ac.uk

Camilla Harris: *LEAF, NAC, Stoneleigh, Warwickshire, CV8 2LZ, UK.*
Email: camilla.harris@leafuk.org

Francis Harris: *School of Earth Sciences and Geography, Kingston University, Kingston upon Thames, KT1 2EE, UK.*
Email: f.harris@kingston.ac.uk

Phil Harris: *Department of Geography, Environment and Disaster Management, Coventry University, Coventry, CV1 5FB, UK.*
Email: p.harris@coventry.ac.uk

Julie Ingram: *Countryside and Community Research Institute, Dunholme Villa, Park Campus, Cheltenham, GL50 2RH, UK.*
Email: jingram@glos.ac.uk

Brian R. Johnson: *Environmental Consultant, 1 Waverley, Stockland, Honiton, Devon, EX14 9BT, UK.*
Email: b.johnson@btinternet.com

James Jones: *Royal Agricultural College, Cirencester, GL7 6JS, UK.*
Email: James.Jones@rac.ac.uk

Raymond A. Jussaume Jr: *Department of Community and Rural Sociology, Washington State University, Pullman, WA 99164-4006, USA.*
Email: rajussaume@wsu.edu

Karlheinz Knickel: *Institute for Rural Development Research, J.W. Goethe-University of Frankfurt/Main Zeppelinallee 31, 60325 Frankfurt/Main, Germany.*
Email: knickel@ifls.de

Fergus Lyon: *Middlesex University Business School, The Burroughs, Hendon, London, NW4 4BT, UK.*
Email: F.Lyon@mdx.ac.uk

Tom MacMillan: *Food Ethics Council, 39-41 Surrey Street, Brighton, BN1 3PB, UK.*
E-mail: tom@foodethicscouncil.org

Ben Mepham: *Centre for Applied Bioethics, School of Biosciences, University of Nottingham, Sutton Bonington Campus, Loughborough, LE12 5RD, UK.*
Email: ben.mepham@nottingham.ac.uk

I. Milla: *Escuela Técnica Superior de Ingenieros Agrónomos, Universidad Politécnica de Valencia, Camino de Vera S/n. 46022, Valencia, Spain.*
Email: etsia@upvnet.upv.es

Gareth Morgan: *The Royal Society for the Protection of Birds, The Lodge, Potton Road, Sandy, Bedfordshire, SG19 2DL, UK.*
Email: Gareth.Morgan@rspb.org.uk

Carol Morris: *School of Geography, University of Nottingham, University Park, Nottingham, NG7 2RD, UK.*
Email: carol.morris@nottingham.ac.uk

Tim O'Riordan: *CSERGE, School of Environmental Sciences, University of East Anglia, Norwich, NR4 7TJ, UK.*
Email: t.oriordan@uea.ac.uk

Sarah Peter: *Institute for Rural Development Research, J.W. Goethe-University of Frankfurt/Main Zeppelinallee 31, 60325 Frankfurt/Main, Germany.*
Email: sarahp@stud.uni-frankfurt.de

Clive Potter: *Centre for Environmental Policy, Imperial College London, South Kensington Campus, London, SW7 2AZ, UK.*
Email: c.potter@imperial.ac.uk

Matthew Reed: *Countryside and Community Research Institute, Dunholme Villa Park Campus, Cheltenham, Gloucestershire, GL50 2RH, UK.*
Email: mreed@glos.ac.uk

Christine Reid: *Natural England, Northminster House, Peterborough, PE1 1UA, UK.*
Email: Chris.Reid@naturalengland.org.uk

Theresa Selfa: *Department of Sociology, Social Work and Anthropology, Kansas State University, Manhattan, KS 66506, USA.*
Email: tselfa@ksu.edu

Susanne Seymour: *School of Geography, University of Nottingham, University Park, Nottingham, NG7 2RD, UK.*
Email: susanne.seymour@nottingham.ac.uk

Michael Steven: *School of Geography, University of Nottingham, University Park, Nottingham, NG7 2RD, UK.*
Email: michael.steven@nottingham.ac.uk

Chris Stoate: *Game and Wildlife Conservation Trust, The Allerton Project, Loddington House, Loddington, Leicestershire, LE7 9XE, UK.*
Email: cstoate@gct.org.uk

Charles Watkins: *School of Geography, University of Nottingham, University Park, Nottingham, NG7 2RD, UK.*
Email: charles.watkins@nottingham.ac.uk

Adrian Williams: *Natural Resources Management Institute, Natural Resources Department, Cranfield University, Cranfield, Bedfordshire, MK43 0AL, UK.*
Email: adrian.williams@cranfield.ac.uk

List of Figures

List of Tables

List of Abbreviations

CAP	Common Agricultural Policy
CEC	Commission of the European Communities
Defra	Department for Environment, Food and Rural Affairs
EU	European Union
FAO	Food and Agriculture Organization
IFM	Integrated Farm Management
LEDC	Less Economically Developed Country
MTR	Mid-Term Review
MEDC	More Economically Developed Country
NFU	National Farmers Union
NGO	Non-Governmental Organization
SP	Single Payment
SSSI	Site of Special Scientific Interest
WCED	World Commission on Environment and Development
WTO	World Trade Organization

Preface

The origins of this collection are in an ESRC Transdisciplinary Seminar Series exploring new approaches to sustainable farmland management which ran in 2004-5. The volume seeks to share with a wider audience some of the research reported on, discussed and debated over the course of 2 years, as well as to introduce emerging work in this field from Europe and North America. The editors would like to thank the ESRC for their support of the series (Project code RES-496-26-0030) and all seminar participants - presenters, discussants, facilitators, attendees - for helping to make the process a success. We would also like to thank Paul Elliott and staff at CABI for their technical assistance and our editors Rachel Cutts and Kate Hill for their help and patience in the production of this volume.

Chapter 1

Agendas for sustainable farmland management[1]

R. Fish, S. Seymour, C. Watkins and M. Steven

Introduction

It is widely recognized that the idea of sustainability has diverse political and cultural expressions. Competing and even contradictory accounts of society–nature relations are frequently marshalled together under its guise. Rendering this category stable in its meaning seems at times a problematic preoccupation, incompatible with how sustainability functions within and across different arenas of policy discourse. Indeed, while the currency and legitimacy of sustainability often turns on the vexed issue of conceptual and practical 'prescription', it is also the case that the idea travels with remarkable efficiency precisely because it allows society to project and secure the future in different ways. Thus the task of critical inquiry into sustainability tends to point in two quite different directions. On the one hand it seeks to iron out diverse and rapidly proliferating agendas into a coherent political and cultural project with operational targets and measurable ends. On the other, it seeks to unpack and interrogate the different meanings that sustainability agendas produce and to consider their implications for practice.

The process under inspection in this book - farmland management - is instructive in both these respects, for, as we shall see, in the occupancy and experience of an ostensibly straightforward material entity like 'farmland', competing expectations, agendas and values are gathering and responding in different ways to the exigencies and moralities of sustainability. Drawing on the insights of researchers and practitioners from different scientific, social scientific and humanities backgrounds and reflecting upon both 'on the ground' experience and 'at the table' developments in the policy arena, this volume reveals sustainability as a key motif of emerging agendas for agriculture, yet one invested with a range of meanings and purposes.

As we shall see, calls for a more sustainable agriculture take place against the backdrop of a wider debate concerning the social and economic restructuring of rural areas in More Economically Developed Countries (MEDCs). While the pace and extent of this restructuring process is frequently disputed, the implication is that the meaning and purpose of contemporary countryside environments, and the role of agriculture within them, is gradually being recast (Essex *et al.,* 2005). To a significant extent these changes have to be understood within the context of a rapidly changing political economy. In the UK for example, World Trade Organization (WTO) reforms and the widening of the EU have questioned levels of farm financial support while a decline in the relative economic importance of agricultural production in rural areas, as industries such as tourism have grown, has challenged agriculturally-defined views of the countryside (Winter, 2003a). Moreover, growing recognition of the negative environmental impacts of farming regimes, such as diffuse pollution, as well as the wider real or perceived threats to human health embodied in food scares and livestock diseases such as foot and mouth and avian flu, have also progressively undermined the historically secure place of farming within agendas for rural economy and society. In response policy reforms at both European and national levels have sought to make agriculture more market-driven, and 'assured' in the areas of environmental protection and animal welfare (e.g. Defra, 2002).

Yet, despite the declining importance of agricultural output to the economies of Western nations, agri-centric views of rural areas and their development continue to assert influence (see Marsden, 2003). In part this is because of the material extent of farmland within rural landscapes: agriculture continues to be the dominant land use in countryside environments and therefore central to their iconography and meaning. They have also endured because new ways of valuing farmland are being developed, particularly those that emphasize agriculture's role as the producer of social, economic and cultural benefits far exceeding conventional commodity production. This idea stands as an important thread to debates surrounding the meaning and centrality of sustainable agricultural systems. Thus, through ideas such as 'multifunctionality', 'joint production' (Marsden, 2003; Wilson, 2007) and most recently 'ecosystem goods and services' (Millennium Assessment, 2005), the so called 'non-market' benefits of farmland environments are being brought to the fore as a means of grasping the wider condition of rural livelihoods and landscapes and how pathways towards sustainability can be described, observed and exemplified in the context of agricultural systems.

Such new, or at least more explicit, ways of valuing agriculture run to the heart of policy reforms at both European and national levels and are indexed in a number of ways, including measures to maintain and enhance valued landscapes and habitats through agri-environmental and land management schemes, as well as the development of novel recreational and leisure opportunities rooted in the use of agricultural landscapes. While these developments have been widely documented and accounted for since the late 1980s, recent theorizations represent an important development on arguments claiming that an era of expanding commodity production is now simply over. In particular they represent an important advance on debates surrounding the transition to so-called 'post-productivist' agricultural systems, a nomenclature which originated in the UK in the early 1990s and where strategies of market survival and re-adjustment implied a crude and wholesale inversion of productivist agricultural logic (Evans *et al.*, 2002; Wilson, 2007). Many researchers have, for instance, critically interrogated the extent to which rural/agricultural landscapes, and the attitudes and value systems that shape them, can be seen to be changing in neat uni-linear terms, either across and within different nations, or at the more immediate level of locale, landscape and farm holding (Wilson, 2001; Marsden, 2003; Walford, 2003). One aspect of more subtle conceptualizations of rural change is that the uneven geography of restructuring processes has been increasingly asserted. Rural areas in the UK have, for example, been framed as 'differentiated countryside', characterized by differing actor spaces and processes (Lowe *et al.*, 2003; Marsden, 2003) where the status of agricultural systems as a viable part of the rural future are marked out quite differently.

All of the contributions that make up this volume are framed, directly or tacitly, by the terms of these restructuring processes, and how to marshal a place for agriculture within them. It is therefore worth acknowledging from the outset that they must be set alongside markedly different ways of imagining rural nature. For instance, in MEDCs where agriculture has been extensively restructured and industrialized in the post-war period, such as the Netherlands and the UK, experiments in the '(re-)wilding' of landscapes via ecological and wilderness theory (Vera, 2000; Parkes, 2006) have been established to explore the environmental implications of withdrawing farming entirely. Such visions serve to challenge prevailing wisdoms as to the extent to which discourses of sustainability and rural land management must be automatically routed through an agri-centric understanding of rural areas. These alternative visions are, of course, subject to their own critiques. In less densely populated rural areas (such as some regions of France), or less deeply industrialized agricultural systems (such as in upland areas of Italy and Greece), a retreat from agriculture is a well-established story of 'abandonment', whose environmental and social outcomes do not sit comfortably with the idea of sustainability at all, not least because there is a palpable risk that communities of knowledge - highly relevant to the idea of sustainable land management - are being consigned to history (Saratsi, 2003; Cevasco, 2005; 2008). In consequence while this volume is primarily concerned with understanding how different pathways towards sustainability may be enabled through iterations of farmland management, it does not follow that debates about land management in rural areas are confined to these logics alone. Much work still needs to be done *not only* on how discourses of sustainability are being coupled with different agricultural systems and regimes, but how these couplings may be complemented, and contested by, quite different assumptions about the meaning and purpose of rural land.

Sustainable development, farmland management and neoliberalism

Sustainability discourses find their most marked and prominent expression in the concept of 'sustainable development', which emerged from the 1970s through the UN-led process of international environment and development discussions, most notably in the 1987 World Commission on Environment and Development (WCED) report, *Our Common Future*. It has been well documented that the Commission set out a framework for understanding sustainable development as a process encompassing environmental, economic and social concerns through which previously opposing discourses of environment and development could then be reconciled (Baker, 2006). Importantly it also foregrounded the goal of meeting human needs within 'not absolute limits but limitations imposed by the present state of technology and social organisation on environmental resources and by the ability of the biosphere to absorb the effects of human activities' (WCED, 1987: 8, cited in McManus, 2000: 813).

Since the 1992 Rio conference sustainable development has become a discursive norm in international circles, with the UN process stimulating a number of international agreements and national policies inspired by its ethos, including sectoral policies relating to agriculture. This is not, however, to suggest there is universal agreement over the term. Rather, Baker (2006: 27) argues that sustainable development is being mobilized as a broad social concept in which 'there is both a readily understood "first-level meaning" and general political acceptance, but around a given set of core ideas lies a deeper contestation'. Indeed Dryzek (2005) argues, drawing parallels between the concept of sustainable development and that of democracy, that such contestations are central to the term's interest and signal its importance. A key area of dispute is over whether the idea of sustainable development set in train by WCED has set out a more or less radical agenda. Critiques of the WCED reveal, for instance, how its key concerns with 'needs', 'limitations' and more open governance offer differing possible interpretations of the idea's political progressiveness. One line of argument is that WCED represented a radical social democratic agenda for change based on a common concern to meet human needs within environmental limits and through more egalitarian mechanisms, which has been subverted by the uneven power relations of the international political and economic system (Baker, 2006). Alternative views regard the process as sympathetic from the outset to neoliberal versions of internationalization and globalization (Whitehead, 2007).

Nonetheless what is clear is that ideas of sustainable development are complexly interwoven with the resurgence of liberal economic regimes. According to Peck and Tickell (2002: 380, 381), neoliberalism works as 'a kind of operating framework' for globalization based on 'free trade, flexible labor, and active individualism'. While the influence of neoliberalism is regarded as pervasive if not hegemonic (Potter and Tilzey, 2007), it has been recognized as both temporally and geographically heterogeneous. Discussions of agricultural change by Lockie and Higgins (2007) draw on this work by Peck and Tickell (2002) to highlight two different styles of neoliberalism: a 'roll-back' style, involving the deregulation and dismantling of the infrastructure of Keynesian styles of governing, characteristic of the 1980s, and a 'roll-out' style, involving re-regulation and 'Third Way' solutions to address the key shortcomings of the 'roll-back' era, characterized by hybrid styles of technocratic and interventionist neoliberal governance, such as partnerships and self-governance (Higgins and Lawrence, 2005a), more typical of the 1990s. Likewise the geographical differentiation of neoliberalism is apparent in practice, with the emergence of diverse, context-specific forms embedded within broader neoliberal frames (Peck and Tickell, 2002). Thus, a key challenge to the formation of sustainable rural futures and agriculture's place within them is the extent to which sustainable farmland management can be reconciled with these economic processes and it is to them that we now turn.

'Roll-back' neoliberalism and agricultural change

As a version of 'roll-back' neoliberalism, the political economy of agriculture can best be understood in agri-industrial terms, characterized by a productivist outlook, open markets, low value bulk commodity production, economies of scale and increased integration into broader agri-food networks (Marsden, 2003; Potter and Tilzey, 2005). The key venue for promoting and policing this mode of production is the WTO, against and through which particular versions of sustainable agriculture are now developing. Countries such as Australia have embraced this 'competitive productivist' form of neoliberalism (Dibden and Cocklin, 2005: 136), while in Europe Potter and Tilzey (2005) argue that there has been increased support for neoliberalization, despite competition from neomercantilist and multifunctionality discourses. They relate this to 'the emergence of *nonproductive* fractions of agro-food capital such as processors, distributors and retailers as key and influential players in the last 20 years', seeking to minimize governance and aligned with WTO as well as government concerns to reduce expenditure and rely more on the market (Potter and Tilzey, 2005: 587). These interests are 'sufficiently disembedded from national and regional contexts and geared to the supply of world markets to be described as global in outlook and orientation' (*ibid.*: 589).

In this context sustainable agriculture has frequently been understood in terms of strong and continuing state support for agricultural production. Central to this argument is the notion that the state should continue to underwrite the productive capacity of agriculture because agricultural systems 'jointly' produce a range of sustainability outcomes which are central to wider rural development goals but which may not be adequately secured through unfettered free trade. In other words, continuing state support for agriculture works to address social objectives (such as retaining farm families on the land) and to protect and enhance landscape, biodiversity and heritage aspects of farmland, providing benefits for the wider public and entrepreneurial opportunities for farmers (Hodge, 2000; Harvey, 2003; Potter and Tilzey, 2005). The European Union is an important exemplar of this process and the tensions that come with it. In this case, the process of defending a 'European model of agriculture' (Cardwell, 2004; Shucksmith *et al.,* 2005), through focusing on and foregrounding its 'multifunctional' nature (Van Huylenbroeck and Durand, 2004), has become the stage upon which its sustainability agendas are thought, by some, to be resisting neoliberal discourse (see Hollander, 2004). This is especially the case in relation to more recent and strong styles of multifunctionality discourse which foreground social and environmental aspects and are supported by a wide-ranging alliance of environmental and farmer organizations (Potter and Tilzey, 2005). Even here though Potter and Tilzey (2005: 593) report the playing of a 'three-level game' in which the EU deploys neoliberal arguments internationally and bi-modal approaches at Community level, whilst accommodating different national styles, with member states such as the UK and the Netherlands favouring decoupling and more market-led approaches and others such as France championing multifunctionality (see also Potter and Tilzey, 2007).

Conversely others see the European strategy as a fundamentally protectionist approach, blind to the social and environmental excesses of industrial agriculture, and with highly debatable outcomes for the sustainability of agricultural systems elsewhere (Potter and Burney, 2002; Morgan *et al.,* 2006; Potter and Tilzey, 2007). This critique has been particularly applied to neomercantilist - weak - versions of multifunctionality which have a productivist view of farmers' role and draw on constructs of agricultural exceptionalism to argue for state protection to support both guaranteed production for domestic markets and export potential. Such discourses are a longstanding component of the Common Agricultural Policy (CAP) and support for commodity-based programmes articulated by key north European farming unions represented by the umbrella grouping COPA, 'a consistent opponent of the radical decoupling of CAP support to its members, active in its support for retaining direct aids to farmers but lukewarm in its attitude to an expansion of Pillar 2' (Potter and Tilzey, 2005: 591).

Arguments about the European model resisting neoliberal tendencies and discourses are further undermined by the growing tendency in Europe to cast the sustainability of agriculture in terms of natural/environmental/habitat/ecosystem 'goods and services', which farmers 'prepare' and 'deliver' and publics 'pay for' and 'consume'. The logic of environmental 'goods and services' has been advanced as a framework through which scientific and social scientific disciplines can rationalize for policymakers what farmland is thought and meant to do (or not to do) and whether these outcomes then matter when managing for sustainability (Potschin and Haines-Young, 2006). However, the pragmatic and communicative appeal of this approach has to be squared with the wider political and cultural discourses that have propagated the use of such terminology. As with the tone and concerns of environmental economics in the 1980s (e.g. Pearce *et al.,* 1989) this systemic way of thinking about environmental sustainability is far from benign. It risks embedding neoliberal ways of thinking about environmental entities and their management, and can function as a means by which other moral registers surrounding the idea of sustainable agriculture are overshadowed and denied (Robertson, 2004; 2006; Potter and Tilzey, 2007). So, for instance, it could be argued that the 'goods and services' logic is part of a process that is progressively understanding the environmental benefits of agricultural production in commodity terms; as awaiting and demanding, disaggregation and pricing by the market. The recasting of water from a common resource to a privatized commodity is a parallel example of this process (see Bakker, 2005) and one which is critically linked to neoliberal strategies of water and agricultural governance in Australia through such market-based instruments as tradable water rights (Dibden and Cocklin, 2005).

We might further say that this is an approach that is in danger of downplaying concerns over the negative and unwanted environmental impacts of agricultural practice and overshadowing the legal responsibilities farmers have to protect their environments. For example, the moral registers implied in such ideas and protocols as 'Duty of Care' in Australia, and 'Good Agricultural and Environmental Condition' or Codes of Good Agricultural Practice in England understand the environment as something more than an instrument of the market, with farmers positioned as stewards or potential law-breakers, and the public as citizens with rights and concerns beyond those of consumers. It is in this sense that farm assurance schemes (such as the 'Red Tractor' initiative in the UK) have been criticized for implying higher-level performance (deserving of a price premium) whilst only achieving legal responsibilities (Marsden, 2003). Likewise, in arenas of environmental protection, such as agricultural soil, water and air pollution, there is a danger that under this discourse of 'goods and services' polluters may be paid from the public purse to clean up pollution, thereby contravening legal duties and the polluter pays principle (witness, for instance the debate over the UK response to nitrate contamination of water - see Seymour *et al.*, 1992). Furthermore recent policy initiatives, such as the EU Water Framework Directive (WFD) (2000), which signal a shift in rural production (and protection) priorities away from food and towards water, serve to strengthen legal responsibilities not to pollute and demand an active restructuring of mainstream agriculture. Such concerns work against a logic of economically bi-modal production forms, albeit that spatial differentiation is being mobilized via measures of environmental risk, embodied in notions such as catchment sensitive farming.

'Roll-out' neoliberalism and agricultural change

While these attempts to commodify nature include attempts to 'roll-back' regulation, they also involve 'roll-out' strategies of re-regulation both by state and non-state actors. The history of sustainable development as reflected in the UN-led work from the 1980s is important here for it has been instrumental in fostering the development of more open structures of governance. A prominent and landmark expression of this is found in the development of Agenda 21 at the 1992 Rio conference with its concern to foster and accommodate much broader and more varied forms of (non-state) knowledge as the basis for practising sustainable development (Baker, 2006). This shift away from direct management by government now features strongly within theorizations of agricultural systems (Higgins and Lawrence, 2005b; Potter and Tilzey, 2005), and stands as an important analytical framework by which we can begin to understand how notions of sustainable farmland management come to be practised and contested. In an era of governance, then, the state tends to act as a facilitator and a moderator of agriculture rather than a direct provider or regulator and, in so doing, the responsibilities of state and non-state actors are increasingly blurred. As an emerging form of 'roll-out' neoliberalism these changes have been the basis for critiquing the tacit politics behind the emergence of sustainable development discourse at the global level. For instance, the increased role ascribed to non-state actors and voluntary actions is thought by many to have cultivated a 'business-oriented notion of sustainable development' by the time of the Rio conference (McManus, 2000: 814), with partnerships between business and other groups being a key governance feature of the most recent 2002 Johannesburg summit (Seyfang, 2003).

Expressions of this governance agenda can also be found at the local level. From the 1990s in particular, there has been a growing tendency to mobilize agricultural sustainability around such discourses of 'community' governance and stewardship as governments seek to avoid more regulation and engage with farmers. This has been a prominent trend in areas where liberal economic regimes have been deployed, such as North America and Australasia, based around initiatives related to forests, watersheds, land conservation and rural development (Warburton, 1998; Seymour, 2004). One of the best known examples is the Australian National Landcare Programme. Though heralded in international circles as 'one of the most significant participatory environmental programs ever developed' (Martin and Ritchie, 1999: 118), it has been subject to wide-ranging critique, principally relating to its failure to stem an ongoing decline in water quality and soil salinization (Dibden and Cocklin, 2005; Lockie and Higgins, 2007). While the focus on collective rather than merely individual action has encouraged landholders to think at the landscape and catchment scales more relevant to addressing natural resource management issues (Higgins and Lockie, 2002), naïve understandings of community capacities and power relations (Lane and McDonald, 2005) have led to the dominance of agri-centric, productivist interpretations of 'landcare' in which environmental and social understandings have been downplayed or ignored when in conflict with economic imperatives (Martin and Ritchie, 1999; Lockie and Higgins, 2007). Outcomes of this nature point to a wider susceptibility of community-based strategies to 'capture' by established producer or landholding interests and highlight that farmers and other rural publics are far from 'passive recipients' of such governance strategies (Singleton, 2002; Seymour, 2004; Dibden and Cocklin, 2005: 148). While stewardship initiatives targeting individual landholders abound in the UK, collective forms have only recently begun to emerge, with examples including the 'Landcare' initiatives of the Environment Agency, the community forestry projects supported by the Forestry Commission, the Countryside Agency's Norfolk Arable Land Management Initiative (NALMI) and a variety of smaller grassroots or NGO based schemes. At the EU level, the implementation of the WFD is likely to mean that the management of agricultural land-water interactions will be an important arena in which these initiatives develop (Warner, 2007).

Thus, despite the implied association with a broader public interest, these and other variations of the 'roll-out' style of neoliberalism have been cast as 'a *privatization* of governing' (Higgins and Lawrence, 2005a: 5). For some, such strategies represent an abdication of state responsibility, with private companies - from large food corporations to farmers - left to manage food and agricultural processes and land conservation at their will (Martin and Ritchie, 1999; Singleton, 2002). These trends have only been strengthened by growing corporate concentration in the food sector, involving both horizontal and vertical integration and increasing market dominance by a few major retailers (Food Ethics Council, 2005), resulting in what Marsden (2003: 166) has termed a growing 'private-interest' regulation. Accompanying this, there has been a large-scale privatization of agricultural R&D as corporate strength has grown and state agricultural advisory services have been marketized, a notable example being the privatization of ADAS in 1998 (see Ingram, this volume). For others, such as Lockie and Higgins (2007) who draw on ideas of governmentality, 'roll-out' neoliberalism is characterized by a new and careful form of state involvement in sustainability through 'action at a distance', revealing 'a number of subtle and novel ways in which governments are attempting to influence the practices of land managers' (Higgins and Lockie, 2002: 420). 'Action at a distance' is a well recognized phenomenon in relation to economic criteria, such as market prices, new technologies or supermarket contracts (Marsden, 2003) and is cast by Higgins and Lockie (2002: 420) as an '*advanced liberal*' mode of governance. Indeed, the style of action encouraged through such distant governing is one which 'seeks to facilitate the conditions for *entrepreneurial self-governing*' by individual farmers (Higgins and Lawrence, 2005a: 5, italics added) in which social and environmental responsibility is embodied in entrepreneurship (Higgins and Lockie, 2002; Lockie and Higgins, 2007). A corollary of this entrepreneurial self-governance is the expectation that farmers become more financially self-reliant: more market-oriented and less reliant on subsidies. While experience with this style of governance in Australia since the 1990s has revealed significant shortcomings, notably in relation to discrepancies between the production management styles of farmers and state-sponsored planning strategies (Higgins, 2005) and unresolved tradeoffs between individual entrepreneurship and public concerns over social and environmental change (Lockie and Higgins, 2007), new hybrid forms of neoliberal governance have emerged. These have taken the form of direct state and private re-regulation as well as private, voluntary standard setting, through mechanisms such as Environmental Management Systems, Best Management Practices and other quality standards, led by producer and retailer groupings (Dibden and Cocklin, 2005; Lockie and Higgins, 2007). This vulnerability and capacity to transform in response to emergent problems is regarded by Peck and Tickell (2002) as a key characteristic of neoliberalism.

Ecological modernization

Alongside ideas of community governance stand the precepts of ecological modernization. While Dryzek (2005: 177) has highlighted the incompatibility of even modest forms of ecological modernization with 'roll-back' neoliberalism, the approach can more easily be accommodated as a form of hybrid 'roll-out', re-regulatory or third way solutions pointed to above and emerging (unevenly) since the 1990s (McCarthy and Prudham, 2004; Lockie and Higgins, 2007). Though there are different styles of ecological modernization - Hajer (1995) has distinguished 'techno-corporatist' 're-tooling' from wider ranging, more self-critical, 'reflexive' ecological modernization, while Christoff (1996) identifies similar 'weak' and 'strong' versions (cited in Dryzek, 2005: 172-174) - all are characterized by a focus on economic growth through accommodating environmental concerns within prevailing systems of capitalism.

Ecological modernization is based principally on coordinated government promotion of market-based eco-efficiency approaches, such as encouraging the use of more environmentally efficient technologies through environmental taxes and trading schemes and cooperation with business. The underlying assumption here is that financial and environmental imperatives are intrinsically and harmoniously linked. Business gains by saving money, for example through reducing wastes, preventing future costly cleanup expenditure, better working environments for staff and/or finding new markets in eco-efficient technologies or consumer products (Dryzek, 2005). Ecological modernization approaches represent for Marsden (2003: 252) a more effective and inclusionary method of addressing public and consumer concerns over food safety and environmental damage, which have frequently been expressed through a 'hygenic mode of regulation'. This hygienic mode, which he argues emerged principally as a reactive corrective to the 'problems' of agri-industrial agriculture, has mobilized 'highly professionalized and bureaucratized forms of environmental safeguards and instruments', creating a plethora of piecemeal regulation which works against holistic, integrated sustainable solutions (Marsden, 2003: 252).

Marsden (2003: 195; 2004) instead positions ecological modernization as part of an alternative rural development paradigm, centred firmly on a 'revised moral economy' of agriculture, and also informed by wider issues of spatial sensitivity and diversity and agro-ecology. Such an approach, he argues, would facilitate new styles of producer empowerment which could avoid the constraints of large retailer contracts geared to corporate operational concerns as well as state-imposed 'clean-up' and 'restrictive protection and maintenance of the environment' (Marsden, 2003: 155, 247). Under this regime he suggests that agri-environmental policies might be extended to include measures which also promote the production of quality foods and local employment with the effect that 'economies of scope' (after van der Ploeg *et al.*, 2000) would be encouraged to compete with economies of scale (Marsden, 2003: 155, 165-6). Even so, critics argue that the social justice elements of sustainable development are neglected within ecological modernization approaches (Dryzek, 2005). Notably, ecological modernization has been criticized for shifting focus from 'common purpose' to the foregrounding of the individual (Baker, 2006: 76), with an emphasis on the sustainable consumer somewhat passively following market signals rather than a more active form of environmental citizenship.

The sustainable citizen

Constructions of the sustainable citizen are a related and contested feature of sustainability governance. Citizenship can be understood broadly as the membership terms of a political unit, comprising a framework of rights and responsibilities through which people organize their lives as part of a group (Smith, 2000: 83-84; Whitehead, 2007: 188). In this vein Dobson and Bell (2006: 5) highlight different traditions of 'liberal', rights-based citizenship (which have been expanded to include debates over environmental and non-human rights - Seyfang, 2006) and 'republican', responsibility-led forms focused on the common good. Debates which connect private actions (such as travelling by plane) with public consequences (such as air pollution) and work by feminists - 'the personal is political' - challenge the focus in earlier citizenship models on the public sphere (Dobson and Bell, 2006; Seyfang, 2006: 387). This attention to the private sphere has drawn ideas of the sustainable consumer into the debate over sustainable citizenship. While some writers, such as Sagoff (1988), draw a distinction between citizen and consumer identities, others, including Barry (2006: 45), suggest that 'consumption is not simply a private act but also an identity-forming act for collective identity' when practised 'mindfully'. Nonetheless the focus in official sustainable development discourses has been on individual consumer choices, including in the weak forms of ecological modernization currently practised in Europe (see above; Seyfang, 2006).

Targeting of the sustainable consumer is premised on the idea that monetary concerns are the prime motivator for individual action and that by each consumer operating according to self-interest, collective benefit is generated. Inevitably this underplays other motivating forces and private actions beyond consumption and focuses on behavioural rather than attitudinal change, thereby limiting the potential catalyst effects of an enhanced sustainability ethos both in terms of scale and scope (Barry, 2006; Dobson and Bell, 2006). Such mainstream versions contrast with alternative visions of sustainable consumers (such as those put forward by the New Economics Foundation) who act ethically and reflexively (rather than just in response to market signals) to socially-embed economic relationships by reconstructing ideas of wealth, use of non-consumption as well as wise consumption strategies and engagement in collective as well as individual consumption-based action (Barry, 2006; Seyfang, 2006). Reflecting on sustainable citizenship more broadly, both Whitehead (2007) and Barry (2006) distinguish conformist and more radical forms, with Barry (2006: 33) distinguishing 'critical sustainability citizenship' derived from civil society from 'regulatory, compliance, or state-based notions'.

Alternative food networks

One of the ways in which these ideas of sustainable citizenship are being cultivated in relation to agriculture is with regard to the formation of alternative food networks; that is production–consumption chains effectively retreating from the discontents of an expanding and liberalized agri-industrial food economy. In this guise it is more localized, civic, engaged and participatory realms of agricultural production and exchange that come to the fore; ones concerned, as Lyson (2004: 62) puts it, 'with embedding local agricultural and food production into the community'. Here the idea of sustainable agriculture is about the development of shorter, more self-reliant supply chains; ones that inspire new trust and knowledge networks between consumers and producers, and which broker an alternative sense of economy (Bell *et al.*, 2004; Watts *et al.*, 2005). However, there is debate over how alternative such networks are as supermarkets adopt 'local food' marketing, and quality foods depend on trusted quality assurance rewarded by market premiums (Goodman, 2004). A key issue is the distinction between locally-produced foods and foods labeled by locality. The former are more commonly associated with alternative production though they have also been aligned with 'defensive localism' (Winter, 2003b). The latter can be seen as part of a post-Fordist, 'quality' product differentiation strategy creating added value for producers in a generally declining food market and niche products for retailers, although such labeling also includes well-established products which have never been fully subjected to Fordist production principles. Perhaps the most well known 'alternative' food is organics. Yet while early organic production was associated with producers committed to a wider green social movement, there is more recent evidence of the emergence of organic producers motivated mainly by higher margins and the mainstreaming of a 'modern' style of organic production which shares many similarities with other styles of niche production (Ilbery *et al.*, 1999; Reed, this volume).

Scales of sustainability

There is growing recognition of the geographical complexity of sustainability discourse and practice in terms of uneven and differentiated spaces of sustainability and relational scales (Whitehead, 2007). Scale has long been an important component of official discourses of sustainable development, with the UN version of sustainable development placing emphasis on *global* decision-making alongside *local* action. Yet both terms have come under critical scrutiny as spaces and scales of sustainability. The ecological writer Shiva (1993, cited in Dodds, 2000: 116-117) has made a telling critique of 'the global' of the UN conference as representing 'a particular local and parochial interest which has been globalised through the scope of its reach'. Likewise calls for local action, most prominently seen in the Agenda 21 policy and the development of Local Agenda 21 initiatives, have been criticized for embodying a rather limited interpretation of the local as the 'politically designated spaces' of local government (Whitehead, 2007: 192) or for operating within frames of naïve localism (Lane and MacDonald, 2005). The use of the terminology of uneven and differentiated sustainability reveals its strong association with globalization where these terms are common critical descriptors. Yet, the value of '[g]eographical variations within the political and social interpretation of sustainability' is also recognized as they may reflect 'appropriate' tailoring to specific contexts (Whitehead, 2007: 211).

In this vein an increased sense of the 'differentiation' of the countryside (Lowe *et al.*, 2003) and of the interrelations between agricultural and regional economies (Winter, 2003a; Lowe and Ward, 2007; Potter and Tilzey, 2007) is leading to greater emphasis on the spatial specificity of sustainable agriculture. In Britain, the notion of the 'differentiated countryside' helps in understanding new territories of sustainable farmland management in 'preserved', 'contested', 'paternalist' and 'clientelist' contexts (Marsden, 2003: 151). For example Marsden (2003: 143-144, 151) outlines 'the productionist and *paternalist* countrysides' of Yorkshire, Humberside and Lincolnshire, characterized by few areas of environmental designation where 'intensive production, farm concentration and new forms of lease-holding and contract farming have continued despite Macsharry' and where land managers are increasingly beholden to 'private-interest models of regulation and governance ... led by the corporate retailers'. These areas contrast with the '*clientelistic* and poorer upland areas of Western Britain' and 'preserved' countryside areas which Marsden (2003: 145) argues are less likely to engage in liberalized agri-industrial food production but which may be able to exploit quality food markets (through such mechanisms as the EU 1992 Designation of Protected Origin (DPO) regulations) due to 'new forms of comparative advantage' linked to environmental landscape designations and a concerned or wealthy consumer base in the locality or further afield. The preserved and contested countryside areas are also more influenced than elsewhere in governance terms by 'local residential interests' (Marsden, 2003: 152). Thus, while Potter and Tilzey (2005: 2007) champion ideas of bifurcated agricultural territories, others, including Marsden (2003: 188), suggest a more complex pattern, strongly related to regional contexts, informed not only by agricultural economies of scale but also of 'scope' and 'synergy' with human and non-human environments.

Sustainable farmland knowledges and transdisciplinarity

A key area of complexity and debate, and one closely connected to the issues raised above, concerns the way ideas of sustainable agriculture are shaped and contested around different concepts and languages of knowledge. As previously argued, accommodating different ways of knowing the world is a common marker of sustainability. In one sense this process of accommodation is about developing new transdisciplinary vocabularies and practices that can allow different constituencies and styles of knowledge - from the academic discourses of the humanities, the social sciences and the natural sciences to the work and views of policymakers, farmers and other publics - to communicate over how to manage for sustainability. To what extent this is possible is a matter of debate for, as we have already cautioned, there are no value-free conceptions of sustainable agriculture to be found, however prosaic, pragmatic and objective they seem to be (Food Ethics Council, 2004). Interlinkages between knowledge and power formations are persistent.

Over the past few decades, knowledge has increasingly been inspected and differentiated through attention to different knowledge styles or varying cultural contexts for knowledge exchange and legitimacy. Debates have been pursued in relation to the nature of scientific knowledge, especially in terms of its engagement with wider social influence and its boundaries with other styles of knowledge, and with respect to the nature and value of 'non-scientific' styles of knowledge, including social science. Varying approaches to the nature of scientific knowledge endure. Some, including prominent scientists and institutions concerned with science and ethics, continue to make claims about the neutrality of scientific research and ascribe controversy over scientific knowledge to a lack of public and social scientific understanding of science (see Gieryn, 1999; Food Ethics Council, 2004). Others, often drawing on sociology of scientific knowledge approaches, argue that scientific practice cannot be disentangled from social processes. Research agendas, methods, review and dissemination are negotiated in relation to business, government and wider social expectations, scientific paradigms and information channels (Latour, 1987; MacMillan, this volume). In this vein, Defra's agricultural science policy of the early 2000s has been criticized as dominated by concerns with economic competitiveness rather than sustainable development (Food Ethics Council, 2004: 45).

Arguments from the sociology of science which position science as a social practice and which identify non-scientific knowledge as relevant to the issue in hand, have helped open sustainability debates to a plethora of knowledges. There is increased interest in different forms of knowledge and expertise and different styles of learning, with a greater overall legitimacy ascribed to 'non-scientific' knowledges (Wynne, 1996; Fish *et al.*, 2003), particularly as the insights of conventional science-policy processes have been shown to be deficient, not least *vis a vis* the expertise of a wider citizenry, including farmers (see Irwin, 1995; Wynne, 1996). In consequence, it is often claimed that one of the pre-conditions and markers of a sustainable agriculture is about contextualizing understandings, tools and approaches in relation to these less official knowledges. How such knowledges are made amenable to wider policy and scientific debates is an interesting question for they tend to invite more interpretive, situated and messy engagements with the practice of sustainable agriculture.

A number of authors have sought to differentiate between knowledge types which are often presented as distinct and opposing styles. Key examples deployed in the agricultural context include: codified and tacit; universal and contextual; testable and experiential; expert and lay; science and social science; know-why and know-how. Yet criticisms of dualistic ways of thinking about knowledge have been made as they play down interactions between knowledge forms and the power relations of their production (see Wynne, 1996; Tsouvalis *et al.*, 2000). For example it is clear that farmers often use different combinations of tacit *and* codified knowledge to help address the messy, situated contexts of farmland which pure science may gloss over (Morgan and Murdoch, 2000; Lyon and Harris, this volume). Recognition of the limits of dualistic thinking however, is matched by concerns over indiscriminate acceptance of different knowledge forms, and there have been a number of calls to clarify their frames of meaning and legitimacy (Collins and Evans, 2002). In particular, concerns have been expressed when science is presented as just another knowledge (Fish *et al.*, 2003). Collins and Evans (2002) argue that the debate over the use of knowledge in decision-making has been obfuscated by a lack of discrimination between rights based on technical expertise and rights based on stakeholder or citizenship status. In making these arguments they identify different types of expertise based on *contextually specific experience*: 'contributory expertise', a characteristic of those able to contribute to the science of the field; 'interactional expertise' which defines those able to interact with participants and study the process of science creation; 'referred expertise' which characterizes those able to understand contributions to a field; and those with 'no special expertise' (*ibid*.: 249). While the first group involves those without certified expertise, for example farmers, the third group includes scientists with no specific knowledge of the area of science in question, thus breaking down the conventional barrier between scientific and non-scientific expertise. Likewise, the identification of 'interactional expertise' creates a space for those with sociological expertise. Collins and Evans (2002: 281) argue furthermore that the key in belonging to a 'core-set community in which expertise is used to adjudicate between competing knowledge-claims and to determine the content of knowledge' is the contributory expertise of those with direct experience of an issue. While useful in highlighting the need to incorporate uncertified expertise, Collins and Evans's (2002) framework has been criticized for its 'naïve' distinction of 'scientific' and 'political' phases of decision-making and its limited view of the role and value of citizen knowledges and participation, particularly in assessing issue framing and helping generate and popularize 'reliable public knowledge' through the deployment of 'civic epistemology' (Jasanoff, 2003: 394, 398; Wynne, 2003). The approach of Collins and Evans (2002) is grounded in scientism which assumes 'epistemological commitment to precision and control' whilst dismissing 'realism and comprehensiveness' as guiding research principles (Wynne, 2003: 406), and neglects consideration of how 'relevant' expert research groupings are generated.

One attempt to address 'the fluid and interactive nature of different ways of sense-making' and the power relations embodied in them, is through the use of a 'knowledge-culture' approach (Tsouvalis *et al*., 2000: 912). Tsouvalis and colleagues (2000: 912) argue that such knowledge-cultures 'have their own rules as to what counts as legitimate knowledge' and embody a 'knowing from within' which combines different forms of knowledge in everyday practice norms. This approach helps generate understanding of the ways in which particular groups of actors become associated with one another and particular styles of knowledge in a 'knowledge-culture', and separated from others (Tsouvalis *et al*., 2000; Morris, 2006; Riley, 2007). Other work on the Agricultural Knowledge System (AKS), which emerges from an agricultural extension tradition, similarly highlights the importance of 'knowledge communities' in which learning and understanding is enhanced (Roling and Wagemakers, 2000). Studies which draw on ideas of knowledge styles and cultures provide valuable insights into how different actor groups regard one another, for example, the tendency for government officials to render farmers as 'technicians' with 'know how' skills but not knowledge generators with 'know why' understanding (Wynne 1996; Burgess *et al*., 2000). While the value of farmers' knowledges has been highlighted in much of this work, there has been a tendency to identify them as a knowledge-culture group, distinct from other knowledge-cultures formed around non-farming groups such as agri-environment scheme managers (Morris, 2006) or advisers (Riley, 2007). In contrast, work informed by ideas of farming styles (van der Ploeg, 1993) and 'habitus' (see Shucksmith, 1993), suggests farmers are a more diverse group - including organic, new entrant, non-agricultural, leisure and consumer oriented, un-unionized - with different knowledge networks and implies a questioning of somewhat naïve (and politically charged) ideas of a single coherent farming community.

Such understandings suggest how knowledge legitimacy is often dependent on context. For example, Jasanoff (1990) and Robertson (2004: 375) have highlighted differences between the qualities of regulatory and research science on the one hand and ecological data produced for scientific ends and those produced for circulation 'in the logics of capital and law' on the other. In his work on ecosystem services in America, Robertson (2004; 2006) examines how a particular language of knowledge, in this case of monetary value and market systems, reframes understandings and uses of wetland ecosystems. Through the deployment of 'a classificatory scheme … to abstract all the messy specificities of the object being modified into a set of categories intelligible in the market', Robertson (2006: 372, 373) argues that mitigation banks for ecological services work to create commodities of 'a consistent identity across space and time' which do not cause controversy amongst market protagonists. Controversies that exist over the ecological robustness and selectivity of rapid assessment methods (RAMs) deployed are countered by a focus in practice on less controversial and more easily measurable attributes. However, increasing demands to distinguish multiply-occurring ecosystem services in the context of multifunctionality require more complex scientific discrimination, and though this promises greater market differentiation its increased complexity is destabilizing for capital (Robertson, 2006: 384). This work also highlights how there are both geographies to, and of, sustainable knowledges. The USA has seen the early development of markets and their requisite knowledges for the mobilization of ecosystem services (Robertson, 2004; 2006). The EU is a site where the discourse of multifunctionality has been used as means of resistance (at least initially) to marketization of farmland management. Likewise, geographies of flood risk or species survival limit the universalization of markets for ecosystem goods and services (Robertson, 2006).

The uneven opening up of such new circuits of expertise is intimately connected with new models of governance, themselves inevitably tied up with wider aspects of neoliberalization we described above (Higgins and Lawrence, 2005a). One response has been for actors to compete to defend and promote their knowledge styles. Scientists police the boundaries of science (Gieryn, 1999) while farmers conversely frequently assert and legitimate their knowledge through reference to their long-term management of farmland, their 'creation' of features valued by nature conservationists and publics and discourses of farming as a natural activity (Lowe *et al.*, 1997; Burgess *et al.*, 2000). There is an obvious politics to such knowledge claims - knowledge is never 'just knowledge' as MacMillan argues (this volume) - it is bound up with notions of trust, power and legitimacy. Nonetheless, some protagonists continue to see science as the only legitimate form of knowledge. Extending communities of knowledge and expertise remains controversial and suggests that the idea of sustainable agriculture can never be realized as a matter of simple, objective, proscription.

The book

This edited collection is an exercise in transdisciplinary reflection on sustainable farmland management. It includes contributions from leading academics and emerging researchers from a range of disciplinary backgrounds including economics, agricultural sciences, geography and environmental sciences, as well as practitioners and researchers from NGOs, government agencies and consultancies, who are involved in informing policy and practice on sustainable agricultural landscapes. Some of the chapters work more clearly to academic agendas and styles, while others reflect the more popular, sometimes campaigning, and practical agendas of NGOs and practitioners.

The book begins with a range of chapters which map out the territory of sustainable farmland management in differing spatial, temporal and production contexts. O'Riordan (Chapter 2) provides a challenging yet refreshingly realistic insight into why sustainable agriculture is proving such an elusive concept to mobilize. The tendency, he argues, remains for procrastination in the face of scientific uncertainties, profit-focused societies and significant disjunctures between stakeholder groups, suggesting that a move towards 'less non-sustainable agriculture' (p.27) is more realistic. With upcoming reviews of WTO, EU and UK agricultural policies and ongoing revisions of supermarket operations, there is as he suggests 'much to play for' (p.29). Baginetas (Chapter 3) illustrates an emerging yet frequently neglected concern in the study of sustainable agriculture: the understandings held by farmers themselves. In the context of research on soil quality in Nottinghamshire, UK, he examines both the diversity of farmers' views of sustainable agriculture and important differences in their definitions when compared to those of scientists and policymakers. These differences he suggests need clear recognition before 'mutually agreed goals' (p.39) can be set. Beckett (Chapter 4) picks up this theme of farmer-government interaction by focusing on farming practice and sustainability in an historic context. Thinking about sustainability is not a new preoccupation, although the terminology is different. Using the example of the Midland open field system he identifies the 16th century as a period of ecological sustainability, challenged during the 17th and 18th centuries through processes of land use change and enclosure which also raised questions of social equity. Nonetheless, Beckett's work highlights how, before World War II, farmers' own initiative, sensitive to resource and family sustainability, helped keep a balance between economic, ethical and ecological concerns outside the stresses of wartime. Reducing the capacity for farmers to use their initiative, he suggests, undermines the capacity for sustainable agriculture. Boardman (Chapter 5) conversely highlights the dangers of post-World War II agri-centric thinking with respect to farmland management on the English South Downs. Here soil erosion from increased and more intensive arable farming, while not immediately threatening the continuance of agricultural output, has led to on-farm soil loss and a reduction in downland habitats, and posed threats to the livelihoods of local populations and the environment through significant off-farm impacts, such as muddy floods and sediment transfer. It is clear from this work and Defra's Catchment Sensitive Farming initiative that farmland management must operate with a sensitivity to landscape-scale concerns. The final contribution in this part by Reed (Chapter 6) examines the rise of organic farming, frequently cited as a core approach to sustainable farming. Despite this, the development of UK organic agriculture has been associated more with rising consumer demand for organics rather than with support from government agri-environment programmes which remained limited in relation to organic agriculture until the early 21st century (Rigby *et al.*, 2001). Indeed, Reed argues that informed consumer interest is 'the central innovation of contemporary organic farming' (p.58), placing it in the forefront of the market-oriented behaviour promoted in Defra's (2002) *Strategy for Sustainable Farming and Food.* However, this development is not without controversy. Unpacking organic farming as a combination of 'a social movement, a brand and a regulatory system' (p.58), Reed highlights how tensions have emerged in relation to the sustainability of organics. On the one hand there are concerns over the loss of the organic ethic as large retailers gain more market share and position organics as just one amongst many consumer choices in an ecologically modern form of the commodification of nature.

On the other, debates have broken out over environmental and animal welfare aspects of organic production as organic standards can be seen to accommodate more controversial intensive practices in certain sectors.

The second part on 'Communities of information and knowledge' examines the disputed arena of knowledge claims for sustainable farmland management and considers different styles of knowledge deployed in a range of contexts. The opening chapter by MacMillan (Chapter 7) questions the current scientific knowledge base for sustainable farmland management in a number of ways. Firstly he argues, *contra* Tony Blair, that science is never 'just knowledge' but contains 'values and assumptions' (p.70) which need identification and critical consideration. The corollary of this is that a wider range of non-scientific knowledges have a role to play in the deployment of science, facilitating a broader attention to the justness of knowledge. It must be acknowledged, however, from the experience of our Seminar Series, that while such views were commonly accepted amongst those from social science backgrounds, they remained highly contested by many natural and physical scientists (see Johnson, this volume). The following chapter by Ingram (Chapter 8) focuses on soil and the role of agronomists in providing advice for sustainable soil management. Her nationwide survey of agricultural advisers in England highlighted a range of types of adviser concerned with soil management, with agronomists most likely to influence farmers due to their regular contact over arable operations. Drawing on information from both questionnaires and interviews she suggests that advisers have good levels of knowledge about best management practices for soil, and that there are signs that 'some agronomists' are shifting from 'productivist modes of thinking' (p.90). Nonetheless, there is also evidence that agronomists have difficulties advising in some key areas for sustainable farmland management, such as accounting for nutrients in manure and cultivation practice where they lack 'experience, skills and involvement' (p.89). Ingram therefore concludes that in order to improve sustainable soil management, practical, process-based training is a key requirement for both advisers and farmers. Lyon and Harris (Chapter 9) focus more centrally on relationships between farmers and scientists in organic farming research and their use of different styles of knowledge. They found that all farmers in their survey undertook some form of 'research', deploying a range of knowledge-generating practices from overt field experimentation mirroring scientific studies to more implicit, tacit assessment approaches. Farmer-scientist research collaboration was also common but is hampered by different priorities and an underestimation of farmers' existing research practices. They advocate the extension of participatory research programmes to help generate more effective organic farming knowledges. Finally, Selfa and Jussaume (Chapter 10) return to a theme raised by Reed (this volume), by taking a step down the food chain to investigate consumer attitudes and practices in relation to sustainable farm food production in Washington State, USA. Using a citizen-consumer framework and drawing on an extensive survey of consumers they consider evidence for linkage between demographic characteristics, expressed values and reported behaviour in relation to organic, environmentally friendly and local food purchasing. While their evidence suggests 'environmental attitudes are not uniformly associated with purchasing or eating local, organic and/or environmentally sustainably produced food' (p.119), strong connections between concerns to preserve farmland (an issue promoted actively by local NGOs) or farmers and sustainable food choices suggest a more modest form of consumer-citizenship combined with 'defensive localism' (Winter, 2003b).

The third part focuses more explicitly on the ethics of farmland production and protection, on their connections with debates over sustainability and calls for new ethical frameworks. The first two contributions by Mepham (Chapter 11) and Buller and Morris (Chapter 12) focus specifically on debates over farm animals and sustainability. Mepham suggests the need for 'a new ethical theory' (p.125) to inform human interactions with non-human farm animals. He reviews the use of animals as a food source drawing on a range of utilitarian, legal, animal rights and environmental arguments, in the context of increased understanding of 'the genetic continuity of human and non-human species' (p.125) and animal sentience, the problems of intensive animal production and the practical reality of current populations of human-dependent farm animals providing food sources for many people. Drawing on Rawls' work on ethical social contracts he advocates its extension into an ethical human non-human contract - albeit a notional one - with farm animals. Buller and Morris (Chapter 12) adopt a more explicit focus on the 'problematic ... relationship between farm animals and agricultural sustainability' (p.135) through an analysis of the major discourses linking farm animals to sustainability: farm animals as threats to, vectors of or (more recently) targets of sustainability, principally through welfare considerations. They explore these tensions in the context of key aspects of agricultural sustainability: its anthropocentric nature, its focus on animals as groups rather than as individuals and disputes over needs. The recent welfare-dominated debates which position farm animals as targets of sustainability, they argue, have tended to re-embed dominant modernist positions through an ethic of care which reinforces ideas that farm animals are different and unequal to, as well as dependent on, humans. Combining arguments about individual animal rights and subjectivity and emerging ideas of impure subjects and hybrid networks, they argue for the development of a broader relational ethical framework within a more relational version of sustainability which takes account of the intertwined 'community' (p.144) of human and farm animals. Finally Johnson (Chapter 13) considers ethical debates over agricultural biotechnology. He approaches these from the stance that humans are part of nature and like many other species they modify their habitats to suit their needs, with evidence of this occurring over the long term - sometimes with disastrous results - and from a Darwinian view of the dynamism of non-human nature. Modifications via selective breeding generated little dispute prior to the emergence of transgenic genetic modification, since when moral arguments have been actively used both for and against the process. Drawing on his earlier conceptualization of human-nature relations Johnson is critical of arguments against transgenic genetic modification based on ideas of 'unnaturalness'. Arguing from a position of the moral neutrality of scientific knowledge (contra MacMillan, this volume) for Johnson it is not the technology itself but the way it is deployed which is open to moral dispute. Surprisingly while ethical codes have proliferated in other areas such as medicine and human genetics - often welcomed by industry as they set out 'common boundaries' (p.151) and if agreed by civil society imply customer acceptance - similar codes for plant and animal breeding have been slow to develop. Johnson concludes that the unsustainability of much conventional agricultural production, the moral imperatives of sustainable development and the contention generated by transgenic processes make ethical frameworks for agricultural research and development a pressing requirement.

The fourth part examines a range of key issues surrounding systems and systemic thinking for sustainable farmland management, including the scales of application and measurement of systems approaches and the benefits and drawbacks of different farmland management systems. Stoate (Chapter 14) examines the extent to which environmental, social and economic benefits are compatible when a multifunctional approach is taken to rural land management. Examples are taken from the Allerton Project, a long term case study managed by the Game and Wildlife Conservation Trust (formerly the Game

Conservancy Trust) at Loddington, Leicestershire, UK. In his chapter Stoate emphasizes the importance of 'indigenous knowledge' (p.166) and the need to take account of farmers' cultural values. Drummond and Harris (Chapter 15) assess the benefits of sustainable farming systems and focus on the promotion of Integrated Farm Management systems by the charity Linking Environment and Farming (LEAF). They examine the auditing of sustainable farming systems and demonstrate the importance of such systems for developing trust by consumers in the production of food. Williams (Chapter 16) concentrates on the environmental and economic aspects of sustainable farmland management, taking particular account of impacts at the farm level. Use is made of a whole farm model approach known as the MEASURES framework, which for a given area identifies the most profitable combination of crops and rotation, taking account of soil type, rainfall and other factors, and calculates its environmental burden. The chapter also demonstrates the value of Life Cycle Assessment (LCA) for estimating all pollution emissions associated with a particular farm. Firth, Milla and Harris (Chapter 17) examine the use of a range of agro-ecological, socio-territorial and economic indicators to assess the sustainability of organic farming on three case study farms. Although the indicators suggested some environmental benefits from organic conversion, they implied that the economic sustainability of the farms had either decreased or only increased slightly, possibly due to the short time period chosen for analysis. This part of the book ends with Jones's examination (Chapter 18) at the farm level of the economic costs and benefits of English agri-environmental schemes. His study of three farms demonstrates very clearly the extent of farm level variation in the cost-benefit from participating in such schemes. For example, farms run largely by contractors stood to gain a greater benefit than farms run by farming families. Moreover, he demonstrates that when direct payments were decoupled from production the main beneficiaries were those farms which had previously attracted the highest levels of subsidy.

The purpose of the final part of the volume is to reflect directly upon some of the emerging political and policy frameworks now governing the idea of sustainable farmland management. While all the contributions that make up the first four parts of the book are situated within a political and policy context, and carry with them implications for how policymakers shape farmland management as an object of sustainability, the concern here is to explore how such processes are currently being given distinct modes of political and policy expression. In doing so, the volume interprets these processes in two senses. First, the contribution by Potter (Chapter 19) is designed to situate the book's problematic in the context of the wider political economy of agriculture and, in particular, to remind us that the relationship between sustainability and farmland will find definition as much at the table of world trade negotiations as it will through practical actions taking place at the farm and field level. Here, sustainable farmland management is a function of political discourse, and in particular, a category around which the moralities and disciplines of neoliberal economic models are being asserted and resisted. Potter points to competing responses to this process in national and pan-European debates, and in so doing, offers a cautionary note to those expecting an imminent, and widely agreed upon, set of policy formulas. Second, this part is designed to explore how policy frameworks are developing pathways through emerging sustainability agendas at national and regional levels. Two case studies are offered. The contribution by Morgan and Reid (Chapter 20) surveys how, in the UK, iterations of sustainable farmland management have been framed and evolved, and offers an interesting overview of how policymakers are now grappling with the meta-processes to which Potter (this volume) alludes. They examine the extent to which environmental priorities are reflected in a range of different agricultural policy instruments and objectives and are circumspect about current and likely progress, suggesting tensions between stated domestic government policies for more environmentally sustainable agriculture and those that favour liberalization and EU reform. In the final contribution, Knickel and Peter (Chapter 21), explain how these wider policy discourses are starting to find expression through regional forms of action. They discuss the work of the Regional Action programme in Germany, highlighting how sustainable forms of agriculture come to be fashioned around a sense of locale, one in which the viability of agricultural systems is understood to be tied inextricably to the cross-sectoral production of public goods.

Together these contributions suggest that, while prospects for the sustainable management of farmland come with few guarantees, we can point to an emerging area of inquiry that is beginning to inspect some of its critical and scientific terms of expression. At the end of this volume (Chapter 22) we reflect on how far this might be suggestive of a new transdisciplinary agenda, one that draws on different cognate areas of expertise to test the possibilities and limits of the book's problematic, but one defying neat, and widely agreed upon, policy prescriptions.

Notes

1. This chapter draws in part on the paper by Fish *et al.* (2006), which introduces the special issue of *The Geographical Journal* on Sustainable Farmland Management.

References

Baker, S. (2006) *Sustainable Development*. Routledge, London, UK.

Bakker, K. (2005) Neoliberalizing nature? Market environmentalism in water supply in England and Wales. *Annals of the Association of American Geographers* 95, 542-565.

Barry, J. (2006) Resistance is fertile: from environmental to sustainability citizenship. In: Dobson, A. and Bell, D. (eds) *Environmental Citizenship*. MIT Press, Cambridge, Massachusetts, USA and London, UK, 21-48.

Bell, M., Jarnagin, S., Gregory, P. and Bauer, D. (2004) *Farming For Us All: Practical Agriculture and the Cultivation of Sustainability*. Penn State Press, Pennsylvania, USA.

Burgess, J., Clark, J. and Harrison, C.M. (2000) Knowledges in action: an actor network analysis of a wetland agri-environmental scheme. *Ecological Economics* 35, 119-132.

Cardwell, M. (2004) *The European Model of Agriculture*. Oxford University Press, Oxford, UK.

Cevasco, R. (2005) Nuove risorse per la geografia del turismo rurale: ecologia storica e risorse ambientali nell'Appennino Ligure-Emiliano. *Bollettino Della Societa Geografica Italiana* Series 12, 10(2) 345-375.

Cevasco, R. (2008) *Memoria Verde*. Edizioni Diabasis, Reggio Emilia, Italy.

Collins, H.M. and Evans, R. (2002) The Third Wave of Science Studies: studies of expertise and experience. *Social Studies of Science* 32(2) 235-296.

Defra (2002) *Strategy for Sustainable Farming and Food*. Defra, London, UK.

Dibden, J. and Cocklin, C. (2005) Sustainability and agri-environmental governance. In: Higgins, V. and Lawrence, G. (eds) *Agricultural Governance: Globalization and the New Politics of Regulation*. Routledge, London, UK, 135-152.

Dobson, A. and Bell, D. (2006) Introduction. In: Dobson, A. and Bell, D. (eds) *Environmental Citizenship*. MIT Press, Cambridge Massachusetts, USA and London, UK, 1-17.

Dodds, K. (2000) *Geopolitics in a Changing World*. Pearson, Harlow, UK.

Dryzek, J.S. (2005) *The Politics of the Earth*. Second edition. Oxford University Press, Oxford, UK.

Essex, S., Gilg, A.W., Yarwood, R., with Smithers, J. and Wilson, R. (2005) *Rural Change and Sustainability: Agriculture, the Environment and Communities.* CAB International, Wallingford, UK.

Evans, N., Morris, C. and Winter, M. (2002) A critique of postproductivism as the new orthodoxy. *Progress in Human Geography* 26, 313-332.

Fish, R., Rubiano, J. and Haines-Young, R. (2003) A science of stakeholders? Institutional visions of landscape and sustainability in the management of the Sherwood Natural Area. In: Fry, G. and Palang, H. (eds) *Landscape Interfaces: Cultural Heritage in Changing Landscapes.* Kluwer Press, London, UK, 147-162.

Fish, R., Seymour, S. and Watkins, C. (2006) Sustainable farmland management as political and cultural discourse. *The Geographical Journal* 172(3) 183-189.

Food Ethics Council (2004) *Just Knowledge: Governing Research on Food and Farming.* Food Ethics Council, Brighton, UK.

Food Ethics Council (2005) *Power in the Food System: Understanding Trends and Improving Accountability.* Food Ethics Council, Brighton, UK.

Gieryn, T. (1999) *Cultural Boundaries of Science: Credibility on the Line.* University of Chicago Press, Chicago, USA.

Goodman, D. (2004) Rural Europe redux? Reflections on alternative agro-food networks and paradigm change. *Sociologia Ruralis* 44(1) 3–16.

Harvey, D. (2003) Agri-environmental relationships and multifunctionality. *The World Economy* 26, 705-725.

Higgins, V. (2005) Governing agriculture through the managerial capacities of farmers: the role of calculation. In: Higgins, V. and Lawrence, G. (eds) *Agricultural Governance: Globalization and the New Politics of Regulation.* Routledge, London, UK, 118-132.

Higgins, V. and Lawrence, G. (2005a) Introduction. Globalization and agricultural governance. In: Higgins, V. and Lawrence, G. (eds) *Agricultural Governance: Globalization and the New Politics of Regulation.* Routledge, London, UK, 1-15.

Higgins, V. and Lawrence, G. (eds) (2005b) *Agricultural Governance: Globalization and the New Politics of Regulation.* Routledge, London, UK.

Higgins, V. and Lockie, S. (2002) Re-discovering the social: neo-liberalism and hybrid practices of governing in rural natural resource management. *Journal of Rural Studies* 18, 419-428.

Hodge, I. (2000) Agri-environmental relationships and the choice of policy mechanism. *The World Economy* 23, 257–273.

Hollander, G. (2004) Agricultural trade liberalization, multifunctionality, and sugar in the south Florida landscape. *Geoforum* 25, 299-312.

Ilbery, B., Holloway, L. and Arber, R. (1999) The geography of organic farming in England and Wales in the 1990s. *Tijdschrift voor Economische en Sociale Geografie* 90(3) 285-295.

Irwin, A. (1995) *Citizen Science: A Study of People, Expertise and Sustainable Development* Routledge, London, UK.

Jasanoff, S. (1990) *The Fifth Branch: Science Advisors as Policymakers.* Harvard University Press, Cambridge, Massachusetts, USA.

Jasanoff, S. (2003) Breaking the Waves in Science Studies: Comment on H.M. Collins and Robert Evans, 'The Third Wave of Science Studies'. *Social Studies of Science* 33(3) 389-400.

Lane, M. and McDonald, G. (2005) Community-based environmental planning: operational dilemmas, planning principles and possible remedies. *Journal of Environmental Planning and Management* 48(5) 709-731.

Latour, B. (1987) *Science in Action: How to Follow Scientists and Engineers Through Society.* Harvard University Press, Cambridge, Massachusetts, USA.

Lockie, S. and Higgins, V. (2007) Roll-out neoliberalism and hybrid practices of regulation in Australian agri-environmental governance. *Journal of Rural Studies* 23, 1-11.

Lowe, P., Clark, J., Seymour, S. and Ward, N. (1997) *Moralizing the Environment: Countryside Change, Farming and Pollution.* UCL Press, London, UK.

Lowe, P., Marsden, T., Murdoch, J. and Ward, N. (2003) *The Differentiated Countryside.* Routledge, London, UK.

Lowe, P. and Ward, N. (2007) Sustainable rural economies: some lessons from the English experience. *Sustainable Development* 15, 307-317.
Lyson, T. (2004) *Civic Agriculture: Reconnecting Farm, Food and Community.* Tufts University Press, Lebanon, New Hampshire, USA.
Marsden, T. (2003) *The Condition of Rural Sustainability*. Van Gorcum, Assen, The Netherlands.
Marsden T. (2004) The quest for ecological modernisation: re-spacing rural development and agro-food studies. *Sociologia Ruralis* 44, 129-147.
Martin, P. and Ritchie, H. (1999) Logics of participation: rural environmental governance under neo-liberalism in Australia. *Environmental Politics* 8(2) 117-135.
McCarthy, J. and Prudham, S. (2004) Neoliberal nature and the nature of neoliberalism. *Geoforum* 35, 275-283.
McManus, P. (2000) Sustainable development. In: Johnston, R.J., Gregory, D., Pratt, G. and Watts, M. (eds) *The Dictionary of Human Geography*. Fourth edition. Blackwell, Oxford, UK, 812-816.
Millennium Assessment (2005) *Ecosystems and Human Well-Being: A Framework for Assessment.* Island Press, Washington, DC, USA.
Morgan K., Marsden T., and Murdoch J. (2006) *Worlds of Food: Place, Power and Provenance in the Food Chain*. Oxford University Press, Oxford, UK.
Morgan, K. and Murdoch, J. (2000) Organic vs. conventional agriculture: knowledge, power and innovation in the food chain. *Geoforum* 31(2) 159-173.
Morris, C. (2006) Negotiating the boundary between state-led and farmer approaches to knowing nature: an analysis of UK agri-environment schemes. *Geoforum* 37, 113-127.
Parkes, P. (2006) *Aspirant Ecologies: A Geography of Rewilding in England and Wales*. Unpublished PhD Thesis, University of Nottingham, UK.
Pearce, D., Markandya, A. and Barbier, E. (1989) *Blueprint for a Green Economy*. Earthscan, London, UK.
Peck, J. and Tickell, A. (2002) Neoliberalizing space. *Antipode* 34(3) 380-404.
Potschin, M. and Haines-Young, R. (2006) 'Rio + 10', sustainability science and the landscape agenda. *Landscape and Urban Planning* 75, 162-174.
Potter, C. and Burney, J. (2002) Multifunctionality in the WTO: legitimate non-trade concern or disguised protectionism? *Journal of Rural Studies* 18, 35-47.
Potter, C. and Tilzey, M. (2005) Agricultural policy discourses in the European post-Fordist transition: neoliberalism, neomercantilism and multifunctionality. *Progress in Human Geography* 29(5) 581-600.
Potter, C. and Tilzey, M. (2007) Agricultural multifunctionality, environmental sustainability and the WTO: resistance or accommodation to the neoliberal project for agriculture? *Geoforum* 38, 1290-1303.
Rigby, D., Young, T. and Burton, M. (2001) The development of and prospects for organic farming in the UK. *Food Policy* 26, 599-613.
Riley, M. (2007) Experts in their fields: farmer-expert knowledges and environmentally-friendly practices. *Environment and Planning A* advance online publication doi:10.1068/a39253 [Accessed 2007]. Available from http://www.envplan.com/abstract.cgi?id=a39253.
Robertson, M.M. (2004) The neoliberalization of ecosystem services: wetland mitigation banking and problems in environmental governance. *Geoforum* 35, 361-373.
Robertson, M.M. (2006) The nature that capital can see: science, state, and market in the commodification of ecosystems services. *Environment and Planning D: Society and Space* 24, 367-387.
Roling, N.G. and Wagemakers, M.A.E. (eds) (2000) *Facilitating Sustainable Agriculture.* Cambridge University Press, Cambridge, UK.
Sagoff, M. (1988) *The Economy of the Earth: Philosophy, Law and the Environment*. Cambridge University Press, Cambridge, UK.
Saratsi, E. (2003) *Landscape History and Traditional Management Practices in the Pindos Mountains of Northwest Greece c.1850-2000.* Unpublished PhD Thesis, University of Nottingham, UK.
Seyfang, G. (2003) Environmental mega-conferences – from Stockholm to Johannesburg and beyond. *Global Environmental Change* 13, 223-228.

Seyfang, G. (2006) Ecological citizenship and sustainable consumption: examining local organic networks. *Journal of Rural Studies* 22, 383-395.

Seymour, S. (2004) 'Community'-based strategies for environmental protection in rural areas: towards a new form of participatory rural governance? In: Holloway, L. and Kneafsey, M. (eds) *Geographies of Rural Cultures and Societies*. Ashgate, Aldershot, UK, 214-237.

Seymour, S., Cox, G. and Lowe, P. (1992) Nitrates in water: the politics of the Polluter-Pays-Principle. *Sociologia Ruralis* 32, 82-103.

Shucksmith, M. (1993) Farm household behaviour and the transition to post-productivism. *Journal of Agricultural Economics* 44, 466-478.

Shucksmith, M., Thomson, K. and Roberts, D. (2005) *The CAP and the Regions*. CAB International, Wallingford, UK.

Singleton, S. (2002) Collaborative environmental planning in the American West: the good, the bad and the ugly. *Environmental Politics* 11, 54-75.

Smith, S.J. (2000) Citizenship. In: Johnston, R.J., Gregory, D., Pratt, G. and Watts, M. (eds) *The Dictionary of Human Geography*. Fourth edition. Blackwell, Oxford, UK, 83-84.

Tsouvalis, J., Seymour, S. and Watkins, C. (2000) Exploring knowledge-cultures: precision farming, yield mapping and the expert/farmer interface. *Environment and Planning A* 32, 909-924.

van der Ploeg, J.D. (1993) Rural sociology and the new agrarian question: a perspective from the Netherlands. *Sociologia Ruralis* 33, 240-260.

van der Ploeg, J.D., Renting, H., Brunori, G., Knickel, K., Mannion, J., Marsden, T., de Roest, K., Sevilla-Guzman, E. and Ventura, F. (2000) Rural development: from practices and policies towards theory. *Sociologia Ruralis* 40(4) 391-408.

Van Huylenbroeck, G. and Durand, G. (eds) (2004) *Multifunctional Agriculture: a New Paradigm for European Agriculture and Rural Development*. Ashgate, Aldershot, UK.

Vera, F. (2000) *Grazing Ecology and Forest History*. CAB International, Wallingford, UK.

Walford, N.S. (2003) Productivism is allegedly dead, long live productivism. Evidence of continued productivist attitudes and decision-making in south east England. *Journal of Rural Studies* 19, 491-502.

Warburton, D. (1998) A passionate dialogue: community and sustainable development. In: Warburton, D. (ed.) *Community and Sustainable Development: Participation in the Future*. Earthscan, London, UK, 1-39.

Warner, J. (ed.) (2007) *Multi-stakeholder Platforms for Integrated Water Management*. Ashgate Aldershot, UK.

Watts, D., Ilbery, B. and Maye, D. (2005) Making reconnections in agro-food geography: alternative systems of food provision. *Progress in Human Geography* 29, 22-40.

Whitehead, M. (2007) *Spaces of Sustainability. Geographical Perspectives on the Sustainable Society*. Routledge, London, UK.

Wilson, G.A. (2001) From productivism to post-productivism ... and back again? Exploring the (un)changed natural and mental landscapes of European agriculture. *Transactions of the Institute of British Geographers* 26, 77-102.

Wilson, G.A. (2007) *Multifunctional Agriculture: A Transition Theory Perspective*. CAB International, Wallingford, UK.

Winter, M. (2003a) Responding to the crisis: the policy impact of the Foot-and-Mouth epidemic. *Political Quarterly* 74, 47-56.

Winter, M. (2003b) Embeddedness, the new food economy and defensive localism. *Journal of Rural Studies* 19(1) 23-32.

WCED (1987) *Our Common Future*. Oxford University Press, Oxford, UK.

Wynne, B. (1996) May the sheep safely graze? A reflexive view of the expert-lay knowledge divide. In: Lash, S., Szerszynnski, B. and Wynne, B. (eds) *Risk, Environment and Modernity – Towards a New Ecology*. Sage, London, UK, 44-83.

Wynne, B. (2003) Seasick on the Third Wave? Subverting the hegemony of propositionalism: response to Collins & Evans (2002). *Social Studies of Science* 33(3) 401-417.

Part I

Scene-setting Sustainable Farmland Management

Chapter 2

The elusive quest for sustainable agriculture

T. O'Riordan

Perspectives

Possibly for the first time in human history, humans know that they have the power to undermine the vitality of planetary life support systems, and the knowledge to act not to do so. We live in a unique age, one where the hand of human transformation is getting more powerful, and where the evidence is streaming in of the damage to the resilient fabric of the natural world. Yet we seem incapable of acting to stop the trend. Maybe we cannot do so. We do not appear to have developed the governing arrangements that allow the world to act as a whole entity, to plan for a genuinely sustaining future for the planet and its inhabitants, and to change course over consumption behaviour and values that inhibit the necessary correctives. There is a school of thought, headed by the controversial Jim Lovelock (2006) which believes that the earth is planning her revenge, and that humans will systematically be eliminated from the species mix over the coming millennium. This possible prognosis certainly cannot be ruled out.

Sustainable agriculture is eluding us. Jules Pretty (2002: 58) articulates that in the UK alone, agriculture creates economic costs, with significant social consequences, worth around £1.5 to 2 billion annually, or around £30 for every citizen in the nation. Other studies he cites estimate the externally imposed costs of US agriculture at £13 billion, while globally the consequences for ecosystem services, human health and viable local economies are inestimable, but certainly huge.

Furthermore, Pretty (2002: 171-190) reveals that age-old schemes for conserving agriculture are steadily being dismembered as folk practices are obliterated or forgotten. Fred Pearce (2006: 297-322) documents a similar destructive history for local water conservation practices in many poor nations. Once water was shared, so vitalizing local produce and inspiring human hope. Nowadays, many of these long-standing practices are being removed by a combination of neglect, lack of safeguard and industrial farming and commercial developments. We seem to be losing the last vestiges of sustainable land and water management. *Agri-culture*, Pretty's (2002) word for stewarding the land as a social as well as economic practice, is disappearing off the planetary map.

Sustainable agriculture and sustainable water management maybe did exist, but we could be losing both the practice and the experience. Any return will have to be re-learnt.

The politics of non-sustainable agriculture

The notion of sustainable development is 20 years old. The World Commission on Environment and Development (or Brundtland Commission) reported in 1987, 15 years after the first UN Conference on the Human Environment held in Stockholm in 1972, and as a stage-setter for the UN Conference on Environment and Development assembled in Rio de Janeiro in 1992. In its chapter on sustainable agriculture (WCED, 1987: 144) the Commission recognized the need to place more emphasis on people, not technology; on ecosystem services, not production; and on the long term rather than the present practices. It argued for sensitive land reform, ecologically restrained open trading agreements, land-stewarding incentive schemes and protection of locally appropriate practices in a globalizing world. It called for a holistic approach to agriculture linking global to local and vice versa, coordinated land, forest and water management on a catchment basis, and the embedding of ecological security in all UN agencies and other land management custodial bodies.

Despite these calls - and a huge array of sophisticated reports for the various sustainability orientated non governmental bodies leading up to the 2002 World Summit on Sustainable Development assembled in Johannesburg as possibly the last global convention of its kind - not even the notion of sustainable agriculture is agreed, let alone its practice or organizing and regulating structures. Why is this?

Willing ambivalence

We seem to exalt in being unable to pin down a definition of sustainable agriculture that is shared by those who influence agricultural practices. Organic? Low input-low output? Ecosystem services enhancement? Integrated pest management? Super-efficient tillage? Carbon neutral farm management? None of these has an agreed definition, none has a band of uncritical supporters without detractors, and none stays reliable as a farming definition over any sensible period of time. Maybe there is a willingness to remain ambivalent. It keeps the chattering classes at work, funds research projects endlessly and enables regulators to avoid clear targets and go soft on inspections and penalties.

To be fair, there is no easy way of defining sustainable agriculture. The WCED (1987) recognized the difficulties of local practices for local conditions, the emerging pressures of global food markets and corporations, and the need for some kind of ecological stewardship remit in all policy and regulatory bodies, including those who deal with assistance-aid, and trade-tariffs. Willing ambivalence will remain with any attempts at defining sustainable agriculture for the foreseeable future.

Scientific uncertainty

The most powerful framing interpretation of sustainable agriculture is that it maintains and enhances ecosystem functioning of soil, water, biotic processes and marine biomass. Sadly, scientific evidence is not yet assembled to determine what the resiliences and limits are on soil stewardship, water care, plant removal and restoration, and marine biota continuation. The margins of tolerance are simply unknowable even with holistic and cross-disciplinary scientific approaches. This is because science does not have an agreed definition of health for ecosystems, soil, plants or aquatic regimes. There may be disciplinary certainties to the notion of health, embedded in physics, chemistry, biology, scientology, economics and politics. But the notion of a comprehensive permanence of ecosystem healthiness will never emerge from the scientific tradition even if heroically cross-disciplinary. Healthiness may only remain as an intrinsic notion, based on culture, experiences, 'feel' and tell tale signs of breakdown. But such interpretations do not fit the scientific mind.

The economics of profit, not well-being

Economic valuation seeks to create profit, a proof of benefits exceeding costs, and persistent evidence of progress. Jules Pretty's (2002) notion of agri-culture is an experience of land and water care that links ecological maintenance to human well-being and cultural morals. In essence, agri-culture converts economics to the common weal, i.e. the basis of permanent survival through co-operative and contented endeavour.

There is little point in beating our collective breasts over the inadequacy of economic valuation to address agri-culture. Many have already done this, well summarized by Pretty (2002: 199). What is important here is to examine why economic analysis continues to avoid the consequences of non-sustaining agriculture. Here we turn to the two points above, namely that there is an organized willingness to remain ambiguous, and that even holistic approaches to science cannot guide us. Yet there is also a significant power issue here, namely that vested interest capital seeks to retain its control over non-sustainable agricultural practices. This is why the European Union's Common Agricultural Policy (CAP) and the World Trade Organization (WTO) and the UN Food and Agriculture Organization (FAO) remain impregnable to promoting and financially supporting sustainable agricultural practices.

Yet well-being is becoming recognized as a viable objective for human happiness and reliable livelihoods. The current debate over what are sustainable lifestyles, promoted by the UK Sustainable Consumption Roundtable (2006) and the New Economics Foundation (2004) amongst others, is beginning to take hold. This is not an easy momentum to discern or to predict. Stronger voices are being heard over the work life balance or poverty alleviation, over security of food, shelter and nurture, over protection against crime, violence and ugliness, and in favour of personal self-esteem and human relationships that give deeper meaning and satisfaction. This is perhaps what the Brundtland Commission (WCED, 1987) might have been contemplating when it argued for people as much as technology in the passage to sustainable agriculture.

The well-being framework has plenty of momentum. It will grow in strength, in clarity and in relevance. It needs to address those whose lives are affected by poverty, race, class, disablement and gender. This is beginning to happen. But until the well-being agenda is intimately regarded as relevant and achievable by those who cannot grasp the relevance of sustainability, it will never emerge from being a middle class angst motif.

Towards less non-sustainable agriculture

In 2008 the CAP is due for wholesale review. Whether it commits to a genuinely fundamental review, and in particular, whether it encompasses sustainability practices, remains in doubt for all the reasons mentioned above. At least there is one opportunity for this, and it should not be lost. What needs to be done?

Global perspective, local strength

Agriculture is a global phenomenon, so we need to address the opening up of new food markets along sustainability lines. The WTO and World Food Programme are both lamentably weak on sustainability principles. There is sufficient awareness of these principles along more simplistic lines rather than in relation to whole-scale ecosystem care, and perhaps this book can help encourage the application of integrated sustainability practices. Certainly this is a task needing to be done.

Localism and local production are bound to grow, but not at the expense of sustainable global trade. Nevertheless urban allotments, school organic gardens, local supply chains for supermarkets and other food supply schemes can all be given momentum, so long as there is a commitment to organic production, minimum and safe use of agrichemicals, local employment and promotion of local produce in the leisure and tourist industries (catering, hotels, accommodation generally, school, prison and hospital food supplies). All of this needs conscious, coordinated regulation by the environment and food agencies (not yet in motion but getting there) and appropriate incentives.

Supermarket sustainability practices

The current UK Competition Commission (2006) Groceries Market Inquiry, which includes the four major supermarket chains (Tesco, Sainsbury's, Asda and Morrisons), could enforce sustainable retailing action plans for all food sourcing, delivery, purchase and disposal. This is a feasible objective and it is one that should be pursued. Already the 'big four' are beginning to jostle for their green credentials. But they are still on an early road to full blooded sustainable development action plans.

Pilot catchment management schemes

The onset of the EU Water Framework Directive in 2007 provides a basis for establishing a number of test case sustainable land use schemes for whole river basins in the UK. If these are tied to the Natural Environment and Rural Communities Act (2006), then these pilot schemes should test out the science and economics of ecosystem services functioning and maintenance. This is critical. Only with a range of test cases can we begin to get the science-economics interface right. And such schemes will continually need a massive dose of local involvement via effective participation.

Pillar 3 – sustainable rural communities

Pillar 1 is the EU term for formal farm support, nearly all on a non-sustainable basis. Pillar 2 is the EU term for agri-environment and farm economic diversification schemes, all farm based. Pillar 3 would link the farm to the local community, offset the danger of alienating farmers from local people, link local consumers to food supply chains, and create sustainable rural farms, and open up the routeways for green spaces close to populated areas. This would in turn allow for schemes to create sustainable whole landscapes, to provide ecological slipping stones for species and habitats threatened by climate change, and for offering scope for healthy exercise for all manner of citizens. A report by the Soil Association (2006) showed that organic farming creates 50% more employment per farm and 30% more jobs per ha than non-organic farms. This compares with a loss of 37 farm workers per day for the agricultural sector in the UK. Protection of organic farming and water stewardship methods, as a form of Pillar 3 development aid could avoid loss of wider resources, maintain community structures, limit migration to non-sustainable cities and reduce carbon dioxide and other greenhouse gas emissions. Pillar 3 is a concept that is global in significance. Sustainable agriculture is a routeway to sustainable, permanent and 'happy' rural communities. Both rely on and exalt in each other.

Futures

There is a much to play for if we seriously address sustainable agriculture in forthcoming reviews of EU, WTO and UK food and farming and rural economies transformation. WWF-UK (2006) has made a valuable start here. All the signs are that these shifts will add to non-sustainability. But if we now realize what can be done, in a planet that could exclude us in a few generations' time, then there is hope and purpose for sustainable agriculture to come of age.

References

Competition Commission (2006) *Groceries Market Inquiry*. [Accessed July 2007] Available at http://www.competition-commission.org.uk/inquiries/ref2006/grocery/index.htm

Lovelock, J. (2006) *The Revenge of Gaia*. Penguin Books, London, UK.

New Economics Foundation (2004) *A Well-Being Manifesto*. NEF, London, UK.

Pearce, F. (2006) *When the River Runs Dry: What Happens When Our Water Runs Out?* Eden Project Book, London, UK.

Pretty, J. (2002) *Agri-culture: Reconnecting People, Land and Nature*. Earthscan, London, UK.

Soil Association (2006) *Organic Works: Providing More Jobs Through Organic Farming and Local Food Supply*. Soil Association, Bristol, UK.

UK Sustainable Consumption Roundtable (2006) *I Will If You Will: Towards Sustainable Consumption*. UK Sustainable Development Commission, London, UK.

WCED (1987) *Our Common Future*. Oxford University Press, Oxford, UK.

WWF-UK (2006) *Counting Consumption: CO_2 Emissions, Material Flows and Ecological Footprint of the UK by Region and Devolved Country*. WWF-UK, Godalming, Surrey, UK.

Chapter 3

The contested concept of sustainability in agriculture: an examination of the views of policymakers, scientists and farmers

K.N. Baginetas

Sustainable development and sustainable agriculture

In 1987 the World Commission on Environment and Development (WCED) offered the most widely used definition of sustainable development (Yunlong and Smit, 1994; Smith and McDonald, 1998):

> Humanity has the ability to make development sustainable - to ensure that it meets the needs of the present generation without compromising the ability of future generations to meet their own needs (WCED, 1987: 9).

Since then, the term sustainable development has increasingly become the main focus of every development policy and currently, it could be argued that it is the dominant paradigm guiding development planning (Smith and McDonald, 1998). Agriculture, seen as a method of production and an inherent part of development planning, is linked with the food sector and the overall economy and so it relates directly to the present and future conditions of environments, economies, and societies (Smit and Smithers, 1993). Since the Brundtland Commission's definition of sustainable development, many definitions have been suggested regarding 'sustainable agriculture', each differing in a subtle way and emphasizing different values, priorities and goals (Pretty, 1995a; 1998).

Although it is relatively easy to describe goals for a more sustainable agriculture, it is more difficult to define 'sustainable agriculture' (Pretty, 1995a; 1998). This has led much of the current debate about the nature and potential of sustainable agriculture to be focused on definitions (Francis and Youngberg, 1990). None of them is entirely satisfactory, not only because they mean different things to different people (Gliessman, 1998), but also because the authors producing them have the implicit misleading assumption that it is possible to come up with a single correct definition (Pretty, 1995a; 1998). According to Pretty (1995a) everyone assumes that agriculture must be sustainable but everyone differs in the interpretations of conditions and assumptions under which this can occur.

Nevertheless, Smit and Smithers (1993) argue that there is some consistency in the definitions. The notion of sustainable agriculture is a multi-dimensional concept which includes environmental, economic, and social components. Most researchers agree that food sufficiency, environmental stewardship, socio-economic viability and equity are important ingredients (Smith and McDonald, 1998). They would probably accept that sustainable agriculture has something to do with the use of resources to produce food and fibre in such a way that the natural resource base is not damaged, and that the basic needs of producers and consumers can be met over the long term (Smit and Smithers, 1993). Thus the concept of sustainability consists of three main aspects: (i) ecological sustainability, (ii) economic sustainability and (iii) social sustainability (Pesek, 1994; Yunlong and Smit, 1994; Smith and McDonald, 1998).

However, the definition of sustainable agriculture is still elusive (Bell and Morse, 1999) and producing a single definition, commonly accepted by the three major stakeholders involved in agriculture (farmers, agricultural scientists and policymakers) is one of the main challenges in the realisation of a sustainable agricultural sector. After all, how can you accomplish something that you cannot define? Within this context, it could be argued that the main obstacle in the 'struggle' to obtain a common definition is that one of the three major stakeholders involved in agriculture, the farmers, have not been suitably involved in the effort. Within the literature, instances of farmers actively contributing to the definition debate are very few (Baginetas, 2005). Their views regarding the meaning they ascribe to the concept of 'sustainable agriculture' have not been sufficiently examined. The main aim of this chapter is to rectify this situation through a presentation of farmers' perceptions of the term and a comparison between their perceptions and those of scientists and policymakers. This will allow me to highlight similarities and differences and identify any common ground on which further efforts of finding a common definition can be pursued.

Methodology

The research involved the conduct of semi-structured interviews with 36 farmers (between March and June 2003) in Nottinghamshire (see Baginetas, 2005 for details). The sampling strategy used was theoretical and purposive sampling (Strauss and Corbin, 1998). The material from the interviews and from an interview diary was transcribed as soon as possible after each interview was conducted. Every transcript was read through to identify themes (after Coffrey and Atkinson, 1996) and the analysis involved going through the material thematically to determine codes and categories (McCracken, 1988; Crang, 1997). For respondent validation, 18 of the farmers were recontacted and all agreed to meet again and go through the interview transcripts (between September and October 2003). Of the 36 farmers interviewed, 34 were 'born' into agriculture and only two of them had families that were not directly involved with farming. Of these, 21 had attended an agricultural college and 15 had learned through experience and by working with their fathers and grandfathers on the family farm. Twenty-one of the farmers practised mixed farming, 12 arable farming and three dairy/beef farming. Twenty-nine used conventional methods of farming, six organic and one biodynamic (von Keyserlingk, 1999). In terms of experience, 14 had been farming for 20 to 29 years, 12 for 30 to 39 years, six for 40 to 50 years, three for 10 to 19 years and one for 5 to 9 years. Fifteen had farms of 100-199 ha, 11 of 200-299 ha, six of 300-399 ha, three of more than 400 ha and only one of less than 100 ha.

Understandings of sustainable agriculture

This chapter focuses on the themes emerging from the farmers' responses to the question: 'What is the meaning of the term "sustainable agriculture" to you?' It also compares the responses obtained from the farmers with those of scientists and policymakers, as published within the relevant literature, and the definition of 'sustainable agriculture' that was adopted by the former MAFF (2000).

Farmers' definitions

Figure 3.1 provides a summary of the farmers' responses to the above-mentioned question which fall into three main categories and four lesser ones. Most of the farmers (10 out of 36) considered the term 'sustainable agriculture' to be synonymous with a process of continuation. They believed that practising 'sustainable agriculture' meant to be able to keep on growing crops, the way they have been doing so far, and to continue practising farming in the long term:

> I suppose the ability to keep farming, sensibly, indefinitely without degrading the soils (Farmer 2).
>
> It means keep growing the same crops in a sensible rotation without knackering the soil (Farmer 3).
>
> Em … just growing crops year after year (Farmer 21).

It could be argued that the farmers belonging to this group defined the term from a practical point of view; they understood 'sustainable agriculture' as the process of 'being able to carry on cultivating' (Farmer 4) the crops they have on their farms. In other words, they derived the definition by consulting their everyday farming reality and the practices performed on their farms.

A second group of farmers (8 out of 36) thought the term 'sustainable agriculture' referred to farming in an alternative way to conventional farming, using different practices, such as those used in organic farming, biodynamic farming and environmentally friendly farming:

> Organic farming is probably the best example of sustainable agriculture because the only things we buy into this farm are chicken food and diesel. The rest of it, all the animal feed is produced on the farm to feed back to the cattle and the cattle muck it (Farmer 6 - organic farmer).
>
> By sustainability now what they usually mean is farming the soil without necessarily any inorganic inputs, this is what we tend to think by sustainability, that's what I consider, … one of the clichés is organic farming (Farmer 12).
>
> I've been hearing the government chucking it out on us. Oh, more about the environment, isn't it? I mean, yeah. Give more subsidies for more … environmentally friendly (Farmer 32).

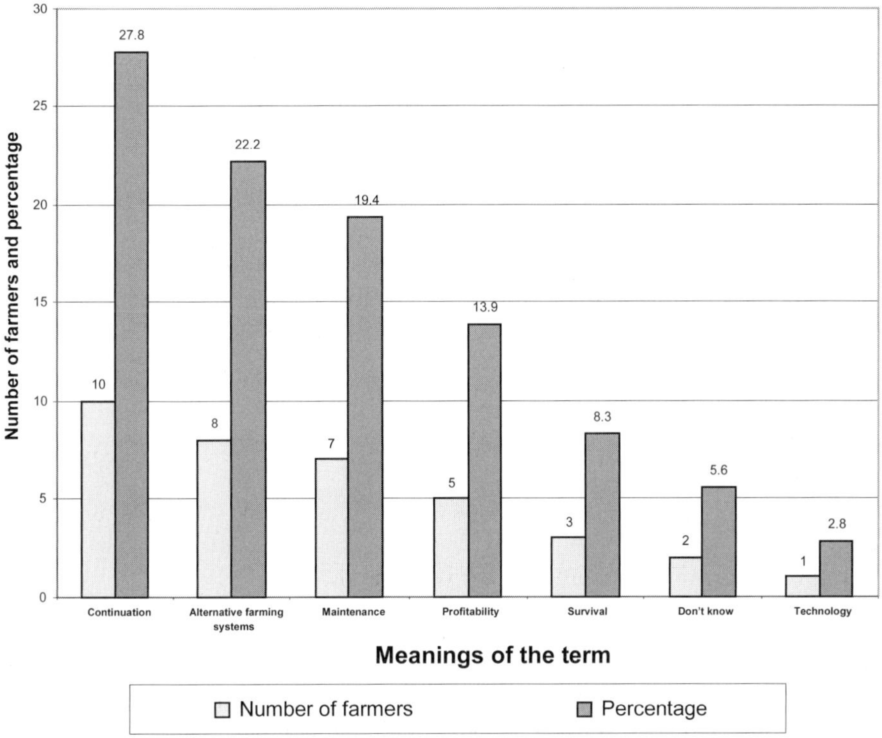

Fig.3.1: Summary of farmers' definitions of the term 'sustainable agriculture'

This group of farmers perceived the term 'sustainable agriculture' to mean something different to the usual practice of farming, unrelated to what they had been considering and practising on their farms. While most of these farmers understood 'usual' as the conventional way of growing crops (which further meant the way that they have been farming so far), they perceived this term to be synonymous to a different way of practising farming, 'sustainable agriculture these days is the organic fad' (Farmer 5); and in search for a name for this set of different farming practices they produced in the interviews the terms 'organic' and 'environmentally friendly' farming.

A third main group of farmers (consisting of 7 out of 36 farmers) believed that 'sustainable agriculture' refers to the process of maintaining their farms. They consider this term to indicate the efforts they should put into sustaining the state of their farms and even the efforts put into improving its state:

> To leave my farm better than I began with it (Farmer 1).
> Just putting back what you take out (Farmer 19 - organic farmer).
> Rotation, putting back things into the soil … fertilizer (Farmer 29).

> Maintaining my business or improving my business … and in your language business being the physical nature of my farm, soils and environment … that's what sustainability means to me (Farmer 34).

These farmers understood the term from the point of view of taking necessary steps so as to be able to 'keep in the fertility to grow' (Farmer 22). They believed that 'sustainable agriculture' meant to apply specific cultivating practices that will allow the essential elements of a farm or a field to replenish themselves at the end of every cultivating period or as one farmer (9) said, 'farming without doing damage'. It could be argued that these farmers associated the term 'sustainable agriculture' with the efforts (which include a set of specific farming practices) of preserving mainly the physical assets of their farms and fields (such as the soil) upon which they depend in order to grow their crops and subsequently produce the optimum yields.

There were also four categories each identified by a smaller number of farmers. There were those (5 out of 36) who believed that sustainable agriculture refers to the profit that you can make out of farming of land and the income that you gain at the end of each farming year:

> Lower input, more output (Farmer 7).
> It means … hem … you are able to get the most for the longest time out of the farm (Farmer 33).
> If I can make a profit (Farmer 36).

It could be argued that these farmers considered the term from a strictly financial point of view. They regarded 'sustainable agriculture' as the 'ability to maintain a fine business' (Farmer 16) which resulted in them securing satisfactory economic returns at the end of each cultivating year. It could be said that this group is an example of farmers that consider themselves as farmer-businessmen who set out every year to achieve, and if possible surpass (in economic terms), their economic goals set out at the beginning of their planning; in an effort not to just be financially viable as a farming family but to be a money-making farming business.

In addition, there were those farmers (3 out of 36) who considered the term to be synonymous with their survival as farmers, associating the term with their continued long-term farming existence:

> Very difficult to explain sustainability but in terms of farming and soil it's more or less thinking about tomorrow as well as of today and making sure that your practices that you perform today aren't going to harm the future and things like that … I mean sustainable to me means you've got to survive. Sustainability to me is survival … basically survival (Farmer 14).

This smaller group of farmers considered the term to be connected with their ability, at the end of the cultivating period or year, to secure their family's living, 'making a living from the land … whether the land will sustain you' (Farmer 20), and also to achieve a satisfactory income that will allow them to continue farming the next year. It could be argued that the reasoning behind these farmers' responses is that they are farming in order to be able to produce enough yields to gain a satisfactory income which will allow them to plan and proceed with the next year's farming preparations.

There were also three farmers who said they didn't know what the term 'sustainable agriculture' meant and one farmer who believed it to be the use of new farming technologies, 'maybe using satellites and technology' (Farmer 27). Furthermore, during the interviews it was obvious that this particular question had an effect on the respondents. After the question had been asked, the farmers' answers (taking time to think and often giving a long, quite descriptive answer) and body language (re-adjusting themselves in their chairs and often grabbing whatever they were holding in their hands more firmly) suggested that the term 'sustainable agriculture' was not one that they frequently used within their community and everyday discussions. As a farmer (14) put it when I asked him the question: 'I mean it's new kind of language isn't it? Of recent times'. This suggests there is a problem of different language and terminology being used between the three different farming actors, namely the farmers, the scientists and the policymakers.

Scientists' definitions

Since the Brundtland Commission's definition of sustainable development in 1987, various agricultural scientists have put a great deal of effort into trying to define 'sustainable agriculture' in precise and absolute terms (Francis and Youngberg, 1990; Altieri, 1995; Ilbery *et al.*, 1997; Gliessman, 1998). This has led to a number of distinct interpretations (Brklacich *et al.*, 1991; Bowler, 1992; Dunlap *et al.*, 1992; Neher, 1992; Yunlong and Smit, 1994; Goldman, 1995; Hansen, 1996; Bowler, 2002a; 2002b). As a result of these attempts, agricultural sustainability has been defined as: a philosophy (Francis and Youngberg, 1990); a set of guidelines for choosing practices (Goldman, 1995); the persistence and the capacity of a system to continue producing for a long time (Fresco and Kroonenberg, 1992); resilience and the ability to bounce back after unexpected difficulties (Conway, 1986); a management strategy (Smith and McDonald, 1998); a production system which does not damage or degrade the natural resources used (Lehman *et al.*, 1993); maximum economic yield (Ainsworth, 1989); economic growth which continues at the same rate or without reducing the natural resource base (Izac and Swift, 1994); and even as an issue of local, national or global proportions, relating to environmental health or food security, now and in the future (Smit and Smithers, 1993; Pretty, 1995b). As Pretty (1995a; 1998) suggests, at least 70 more definitions have been proposed regarding the term 'sustainable agriculture', each differing in a subtle way and emphasizing different values, priorities and goals.

It could be argued that in the case of the definitions provided by the scientists, one encounters the implicit influence that the specific discipline's empirical and methodological assumptions have on the proposed definition. In other words, each scientist attempting to produce a single definition of the term is inherently supporting his or her reasoning on the basis of their discipline's methodological principles. Thus the scientists' attempts result in a multitude of definitions, which is making the drawing of conclusions a difficult task for every researcher.

Policymakers' definitions

Since my research draws on the understandings and experiences of English farmers, for the purpose of this chapter, I decided to use the official definition of 'Sustainable Agriculture' that was provided by MAFF (2000), in order to represent the views of policymakers' within the British farming context.

In 1992, nearly 180 countries met at the 'Earth Summit' in Rio de Janeiro to discuss how to achieve sustainable development. During that meeting a plan of action was agreed upon, Agenda 21, and it was suggested that all countries should produce national sustainable development strategies. The UK was one of the first to do so by publishing *Sustainable Development: The UK Strategy* in 1994 (HMSO). Furthermore, in 1999 the government after an extended consultation process and building on the achievements of the 1994 strategy published a second strategy, *A Better Quality of Life: A Strategy for Sustainable Development for the UK* (DETR). In this second strategy a framework for action was set to deliver sustainable development in the UK. Accepting sustainable development to represent 'a better quality of life for everyone, now and for the generations to come' (DETR, 1999: 8), the UK government proposed four objectives to be met, in order for sustainable development to be achieved:

- Social progress which recognizes the needs of everyone.
- Effective protection of the environment.
- Prudent use of natural resources.
- Maintenance of high and stable levels of economic growth and employment (DETR, 1999: 8).

Following this document, and recognizing that agriculture can play an important part in helping deliver the above strategy and achieving a better quality of life for everyone, now and for generations to come, the UK government produced a third publication, *Towards Sustainable Agriculture. A Pilot Set of Indicators* (MAFF, 2000). In this publication the policymakers acknowledging that 'there is no universally accepted definition of sustainable agriculture' (MAFF, 2000: 5), presented the MAFF definition of 'sustainable agriculture' which 'lies behind the thinking in this document and in all our policies' (MAFF, 2000: 5):

- Ensuring the continuing availability to the consumer of adequate supplies of wholesome, varied and reasonably priced food, produced in accordance with generally accepted environmental and social standards.
- Maintaining a competitive and flexible industry which contributes to an economically viable rural society.
- Ensuring effective protection of the environment and prudent use of natural resources.
- Conserving and enhancing the landscape, wildlife, cultural and archaeological value of agricultural land.
- Respecting a high level of animal welfare (MAFF, 2000: 5).

As can be seen, the policymakers in their attempt to provide a definition of the term 'sustainable agriculture' first break down 'agriculture' to its constitutive parts (food, industry, environment, land and animals) and then focus on each one of these within the sphere of 'sustainability'. Finally, the policymakers propose the above attributes as the characteristics of a 'sustainable agricultural system'.

Discussion

Considering the proposed definitions of the concept of 'sustainable agriculture', it is possible to produce a set of characteristics pertaining to each stakeholder group, according to their views of the concept and as they apply them in their efforts to construct and present their understanding of the term. Table 3.1 presents a summary of these characteristics. Furthermore, based on the above characteristics of the proposed definitions of 'sustainable agriculture' by each group, one could attempt to discover any similarities or differences existing between the groups which inform their attempts to produce a definition of the term. Table 3.2 presents a list of such similarities and differences.

Table 3.1: Summary of the characteristics of the proposed definitions of sustainable agriculture for farmers, scientists and policymakers

Characteristics of the proposed definitions of sustainable agriculture		
Farmers' definitions	**Scientists' definitions**	**Policymakers' definitions**
From a practical point of view	From a general point of view	From an all-encompassing point of view
Based on the everyday practice in their farms	Based on the theoretical underpinnings of their discipline	Based on shedding light to the 'bigger picture'
Priority to their training	Priority to their specialization and professional training	Priority to their scientific advisors' reports
Focus on physical outputs, especially on what they can see and perceive (like yields)	Focus on theoretical or empirical studies and statistical analyses	Focus on consultations and reports produced by scientists
Their farm constitutes their 'world'	They are on the 'periphery' of the farm	They view farming as 'outsiders' (Chambers, 1983)
Their main aim is their self-interest and their family's needs	Their main aim is research publications	Their main aim is the production of documents with measurable targets and deadlines
They use their own 'language'	They use their own terminology	They use their own vocabulary
Main characteristic of given definitions		
Farmers	**Scientists**	**Policymakers**
Practical sustainability, the one that can be seen and measured in their fields and farms	Ideal sustainability, the one that should be achieved by farmers	A broad definition of sustainability, one seen in governmental documents, in compliance with the Minister's directions and/or EU directives and targets

Table 3.2: Group similarities and differences on which the proposed definitions are based

Comparing the influences which inform and affect the three stakeholder groups when attempting to produce a definition of the concept of 'sustainable agriculture'	
Similarities	**Differences**
Based on their own learning and training	Different working environments (farms in contrast to laboratories and offices)
Based on their own concerns and perceptions	Different levels of perception (referring to tangible or intangible concepts)
Based on their own aims and set targets	Different level of involvement (hours of practice)
---	Different level of livelihood impacts (income based on the produced yield in contrast with the research output)
---	Different working motives (production of yield in contrast with production of publications)
---	Different working language

The three stakeholder groups have different interests and thus promote different views of what 'sustainable agriculture' means. Differences in their learning (the practical training of the farmers compared to the discipline based training of the scientists), in their perceptions (farmers understanding only what they see whereas scientists understand the intangible outputs as well) and in their livelihoods (farmers income is based on the produce yields whereas that of the scientists relies on the produced research), affect significantly the views that they have regarding the term 'sustainable agriculture' and its definition.

The analysis confirms that different groups and interests promote different views in order to define and to achieve agricultural sustainability and there is a difference in the interpretations of conditions and assumptions (Pretty, 1995a; 1995b). This supports Gliessman's (1998: 13) suggestion that the term 'sustainable agriculture' is one that 'means different things to different people', and this is made clear from the differences existing between the definitions proposed by the three farming stakeholder groups. Furthermore, people involved in agriculture differ in the environmental, social and economic conditions within which they were brought up and within which they have to live, and it could be argued that these conditions have a strong influence on their attempts to produce a definition of the concept of 'sustainable agriculture'. As Campbell (1994) suggests sustainability, similar to other abstract notions such as beauty, is 'in the eye of the beholder' and it is inevitable that assessments of relative sustainability are socially constructed, which leads to so many different definitions of the term (Pretty, 1995b).

As seen from the results presented above, 'sustainable agriculture' is a value-laden, subjective term with different meanings for different people. This supports Marten's (2001: 121) suggestion that 'perceptions shape the interpretation of information when it enters a social system from an ecosystem, and perceptions shape the decision-making process that leads to actions affecting the ecosystem'. It could even be argued that one person's practice of 'sustainable agriculture' could be another's degradation and exploitation. Given the highly contested nature and definition of 'sustainable agriculture' (Pretty, 1995a; 1998) it is reasonable to suggest that debates surrounding the term should allow all relevant stakeholders to provide their perspective, including practising farmers who have often been neglected in this process. Such discussions could increase the possibilities of producing a definition of 'sustainable agriculture' that has wide acceptance from the three main stakeholders involved in agriculture. A move towards mutually agreed goals is likely to enhance participation within, and subsequent successful implementation of, any agricultural development projects.

Acknowledgements

This chapter was based on research funded by a University of Nottingham Research Scholarship (2001-2004). I would like to express my gratitude to all the farmers who agreed to participate in the research and to my supervisors, Professor Charles Watkins and Dr Richard Field, for their support during the research. I would also like to thank Dr Susanne Seymour for her useful comments and help throughout the above research and during the writing of this chapter.

References

Ainsworth, E. (1989) LISA men have called you. *Farm Journal* 113 (February) 1.

Altieri, M.A. (1995) *Agroecology. The Science of Sustainable Agriculture.* Intermediate Technology Publications. London, UK.

Baginetas, K. (2005) *Agri-environmental sustainability indicators: inter-disciplinary development of soil quality indicators in Nottinghamshire, UK.* Unpublished Ph.D Thesis, School of Geography, University of Nottingham, UK.

Bell, S. and Morse, S. (1999) *Sustainability Indicators. Measuring the Immeasurable?* Earthscan, London, UK.

Bowler, I.R. (1992) Sustainable agriculture as an alternative path of farm business development. In: Bowler, I.R., Bryant, C.R. and Nellis, D. (eds) *Rural Systems in Transition: Agriculture and Environment.* CAB International, Wallingford, UK, 237-253.

Bowler, I.R. (2002a) Developing sustainable agriculture. *Geography* 87, 205-212.

Bowler, I.R. (2002b) Sustainable farming systems. In: Bowler, I.R., Bryant, C.R. and Cocklin, C. (eds) *The Sustainability of Rural Systems: Geographical Interpretations.* Kluwer, Dordrecht, The Netherlands, 169-188.

Brklacich, M., Bryant, C. and Smit, B. (1991) Review and appraisal of concepts of sustainable food production systems. *Environmental Management* 15, 1-14.

Campbell, A. (1994) Participatory inquiry: beyond research and extension in the sustainability era. Paper presented at the International Symposium *Systems-Oriented Research in Agriculture and Rural Development.* Montpellier. 21-25 November 1994, France.

Chambers, R. (1983) *Rural Development: Putting the Last First*. Longman, London, UK.

Coffrey, A. and Atkinson, P. (1996) *Making Sense of Qualitative Data*. Sage, Thousand Oaks, California, USA.

Conway, G.R. (1986) *Agroecosystem Analysis for Research and Development*. Winrock International, Bangkok, Thailand.

Crang, M. (1997) Analysing qualitative materials. In: Flowerdew, R. and Martin, D. (eds) *Methods in Human Geography*. Longman, Harlow, UK, 183-196.

DETR (1999) *A Better Quality of Life. A Strategy for Sustainable Development for the UK*. Department of the Environment, Transport and the Regions. London, UK.

Dunlap, R.E., Beus, C.E., Howell, R.E. and Waud, J. (1992) What is sustainable agriculture? An empirical examination of faculty and farmer definitions. *Journal of Sustainable Agriculture* 3(1) 5-39.

Francis, C.A. and Youngberg, G. (1990) Sustainable agriculture. An overview. In: Francis, C.A., Flora, B.C. and King, L.D. (eds) *Sustainable Agriculture in Temperate Zones*. John Wiley & Sons. New York, USA, 1-23.

Fresco, L.O. and Kroonenberg, S.B. (1992) Time and spatial scales in ecological sustainability. *Land Use Policy* 9, 155-168.

Gliessman, S.R. (1998) *Agroecology. Ecological Processes in Sustainable Agriculture*. Ann Arbor Press, Michigan, USA.

Goldman, A. (1995) Threats to sustainability in African agriculture: searching for appropriate paradigms. *Human Ecology* 23(3) 291-334.

Hansen, J.W. (1996) Is agricultural sustainability a useful concept? *Agricultural Systems* 50, 117-143.

HMSO (1994) *Sustainable Development: The UK Strategy*. HMSO, London, UK.

Ilbery, B., Chiotti, Q. and Rickard, T. (1997) Introduction. In: Ilbery, B., Chiotti, Q. and Rickard, T. (eds) *Agricultural Restructuring and Sustainability. A Geographical Perspective*. CAB International, Wallingford, UK, 1-9.

Izac, A.M. and Swift, M.J. (1994) On agricultural sustainability and its measurement in small-scale farming in Sub-Saharan Africa. *Ecological Economics* 11, 105-125.

Lehman, H., Clark, E.A., and Weise, S.F. (1993) Clarifying the definition of sustainable agriculture. *Journal of Agricultural and Environmental Ethics* 6(2) 127-143.

MAFF (2000) *Towards Sustainable Agriculture. A Pilot Set of Indicators*. Ministry of Agriculture, Fisheries and Food, London, UK.

Marten, G.G. (2001) *Human Ecology. Basic Concepts for Sustainable Development*. Earthscan, London, UK.

McCracken, G.D. (1988) *The Long Interview*. Sage, Newbury Park, California, USA and London, UK.

Neher, D. (1992) Ecological sustainability in agricultural systems: definition and measurement. *Journal of Sustainable Agriculture* 2(3) 51-61.

Pesek, J. (1994) Historical perspective. In: Hatfield, J.L. and Karlen, D.L. (eds) *Sustainable Agriculture Systems*. Lewis Publishers, Florida, USA and London, UK, 1-20.

Pretty, J.N. (1995a) *Regenerating Agriculture. Policies and Practice for Sustainability and Self-reliance*. Earthscan, London, UK.

Pretty, J.N. (1995b) Participatory learning for sustainable agriculture. *World Development* 23(8) 1247-1263.

Pretty, J.N. (1998) *The Living Land. Agriculture, Food and Community Regeneration in Rural Europe*. Earthscan, London, UK.

Smit, B. and Smithers, J. (1993) Sustainable agriculture: interpretations, analyses and prospects. *Canadian Journal of Regional Science* 16(3) 499-524.

Smith, C.S. and McDonald, G.T. (1998) Assessing the sustainability of agriculture at the planning stage. *Journal of Environmental Management* 52, 15-37.

Strauss, A. and Corbin, J. (1998) *Basics of Qualitative Research. Techniques and Procedures for Developing Grounded Theory*. Second edition, Sage, Thousand Oaks, California, USA and London, UK.

von Keyserlingk, A.G. (1999) *Developing Biodynamic Agriculture: Reflections on Early Research.* Temple Lodge, London, UK.

WCED (1987) *Our Common Future*. Oxford University Press, Oxford, UK.

Yunlong, C. and Smit, B. (1994) Sustainability in agriculture: a general review. *Agriculture, Ecosystems and Environment* 49, 299-307.

Chapter 4

Historical dimensions of sustainable farming

J.V. Beckett

Introduction

Much of the sustainability debate is couched in terms of the future; in particular, how will agricultural sustainability be maintained in an environment where population increase is predicted for the period to c.2050 and these predictions suggest that some of the poorer parts of Less Economically Developed Countries (LEDCs) will not be able to maintain existing standards, let alone achieve the aim of feeding their populations. Prediction also implies policy: in other words the point of this exercise is not simply to establish the problem, but to drive policy. In turn, it is hoped this will enable LEDCs to feed more mouths without suffering famine (Sen, 1981; 1986). Is there a historical dimension which might help to guide these debates?

This chapter draws on research, funded by the ESRC between 1999 and 2002, on 'Sustainability in English Agriculture, 1500-2000' (ESRC Data Archive, SN 4517; SN 4537; Turner *et al.*, 2003). The research team comprises historians of land and agriculture, and our particular approach to this subject is not that of scientists, agricultural economists, or geographers. We are interested in how farmers adapted in the past to changes which we would today see as potentially threatening sustainability. But we have also couched our work in such a way as to be able to ask questions about potential lessons from history in relation to agricultural sustainability. In other words, if predictions relating to LEDCs are to be based on an understanding of development in More Economically Developed Countries (MEDCs), then an appreciation of how the farmer handled the problems which we would today describe in terms of sustainability may provide a useful contribution to the wider debate. We have examined the evolution of British, and by that we mean predominantly English agriculture, over a long time span in order to look for periods of change when sustainability came under threat.

There is a view, most succinctly expressed by Conway and Pretty writing of the period prior to 1945 or thereabouts, that:

> agriculture, for most of its history, has been environmentally benign. Even when industrial technology began to have an impact in the eighteenth and nineteenth centuries, agriculture continued to rely on natural ecological processes. Crop residues were incorporated into the soil or fed to livestock, and the manure returned to the land in amounts that could be absorbed and

> utilised. The traditional mixed farm was a closed, stable and sustainable ecological system, generating few external impacts. Since the Second World War this system has disintegrated (Conway and Pretty, 1991: 1).

We would not wish to depart substantially from this viewpoint, but we do question whether such a stark distinction between pre- and post-1945 is likely to overlook the fact that sustainability is not only about ecological considerations, and that many of the economic issues, and questions of equity, which are part of the modern debate have also applied in the past. We may today raise questions about how agriculture can and should develop in a world where population growth can be predicted over the next 50 years, but how did farmers respond to rising population in the past, particularly in a past which lacked economic forecasts and in which there was no Defra, no European Union, and no clear government policy? And, of course, there were no government subsidies.

England is a country of multiple soils and varied terrain, and this has generated numerous farming systems in the past. In the first part of this chapter the emphasis will be on one system and its adaptation. This is the open field farming system which was typical of much of midland England, and which is best recalled in terms of common fields, strips and other 'medieval' features. It survives today in a truncated form in the village of Laxton, near Newark, Nottinghamshire (Beckett, 1989). In terms of agricultural sustainability, from the current literature we have adopted three concepts: ecology, economy and equity (Turner *et al.*, 2003: 235-237). We consider the successful farming community in the past to have been one in which all three are in balance, but that any form of change, from within the farming system or external to it, was likely to have had an impact on one or more of these aspects of sustainability.

The Midland open field system in the 16th to 18th centuries

Although there is still some debate about the origin of the open field system, there is not much doubt that it emerged over several centuries from the later Saxon period in response to the demands of a rising population. Land was brought into cultivation and shared out in the community. Each community needed to be relatively self sufficient in terms of its produce, so that a balance had to be maintained between arable cropping and animals. The system survived the Black Death of the 14th century, and was still in good working order by the time population numbers recovered in the 16th century.

The open field system evolved in such a way as to meet the requirements for ecological sustainability. The throughput of crops in the different rotations (wheat, barley, rye, peas, beans, oats, weeds and grasses) provided good biodiversity, and the introduction of a feed field (the pease field) to provide animal fodder helped to offset the reduction in the area of permanent grass in periods of buoyant grain prices. The system maximized the use of the given gifts of nature through its flexibility. It minimized loss by maintaining ground cover and using a variety of plants, which kept nutrients within the system. Recycling, through the actions of sheep and their natural tendency to tread in their waste, along with the ploughing in of anything left uneaten by the animals, also helped with biodiversity. The principle of farming in nature's image was practised, and the successful integration of plants and animals helped to ensure that the system remained viable across the centuries.

Although it remained viable, it could not and did not remain unchanged, because it necessarily adapted to external change. We now know that in the mid-17th century population growth, which had been steadily upwards since the mid-14th century, reached a plateau and for 80 years or so it was either steady or in decline. Open field farmers came under pressure from forces which they did not understand and over which they had no control, primarily a downwards trend in population. The prices of meat and dairy products held up better in these population conditions than grain, which encouraged farmers to maximize the use of grassland. The Midland claylands were well suited to permanent grass rather than mixed farming. This encouragement was turned into practical help by changing market conditions. Increased quantities of grain were traded, especially from East Anglia, along improved river courses and, eventually, canals. These were ideal conditions for regional specialization (Thirsk, 1984-5; Williamson, 2002).

In response to these pressures the farmers had no predictions of future demand, and nor did they have any help (or hindrance for that matter) from central government beyond the introduction from the late seventeenth century of premiums on grain sales into Europe - which were mostly beneficial to farmers near the ports. Open field farmers responded to these changes in ways which we would today see as raising questions about sustainability. First, they altered the balance between arable and grass, thereby threatening the ecological balance; and second, in making this alteration they affected the equitable distribution of property and rights within the community. How did they make these changes, and what was the impact?

Open field farmers altered the balance between arable and grass by maximizing the use of grass in the village (such as paths and so forth) and then by converting arable to temporary or eventually permanent grass. Since the quantity of land was finite they could not colonize new land, so they had to adapt what they had. Land, either in closes, or in temporary enclosures within the open fields (known as 'leys'), was colonized for temporary grass, which in the case of closes often became permanent. In the open fields the temporary grass leys were put back into arable cultivation, but the period spent under grass clearly had the effect of improving biodiversity and helping to integrate livestock and crops in a balanced, ecologically sustainable manner.

But what might have begun as a temporary solution soon turned into a permanent condition. Land in temporary leys might eventually become permanent grass and through time individual furlongs (groups of strips) and even whole fields were removed from the open field system. The trend continued when the population began to increase again from the mid-18th century because, although the price of grain recovered, the prices of livestock products continued to rise. The final phase was enclosure, with the open fields disappearing altogether to make way for permanent grass. Although best known in its Parliamentary guise (mainly 1760-1830), enclosure in fact took place in different phases and by different means from the 17th century onwards. In turn, a new, profitable, farming system was moulded from an older one, hence Arthur Young's frustration when he travelled through the Vale of Aylesbury in 1771:

> The whole country [from Aylesbury to Buckingham] is open field … the soil [of the Vale] among the richest I ever saw, black putrid clay … As for the landlords, what in the name of wonder is the reason for their not enclosing? All the Vale would make as fine meadows as any in the world (Young, 1771: 18-24).

These changes to the nature of farming affected the equitable distribution of property and rights within the community. The open field system was a communal system in which all the farmers shared the risks and the profits. They operated under the auspices of a manor court, which was the province of the landlord for property transfer, but also the forum in which farmers could debate current practice and take decisions about future farming. Courts tended to be conservative. Large farmers were able to drive the business, while landlords, encouraged by their stewards to assume that a move towards grass would enable their tenants to pay higher rent, were usually happy to encourage change.

The courts tried to regulate what was happening, but gradually they fell into disuse. Enclosure, particularly when sanctioned by Act of Parliament, put them finally into abeyance, extinguished common rights of pasturing, and did away with the last vestiges of communal farming. The tenant farmers may have preferred farming in severalty, but the final adaptation of the system was in effect its collapse as village after village across the Midland Plain abandoned open field mixed farming for permanent grass. This was a major change which significantly altered both the ecological balance and the equitable balance of the village, and it took place under the eye of a benevolent government which facilitated enclosure from around 1760 through private Acts of Parliament. Much of this land came back into arable only after 1939.

When put under pressure in the past farmers responded by trial and error. In terms of sustainability they had to find a new ecological balance in order to make an economic living, without compromising the long term biodiversity of the soil. The major loser was once thought to have been the agricultural labourer forced from the land to become the new industrial worker, but modern research suggests a more complex pattern. The village community was greatly affected in the long run, but since the process of change took decades rather than years the process of adaptation was gradual: the equity of the community may have been upset but it was not necessarily fatally wounded.

The impact of war, 1793-1815

The changes took place predominantly over the period from the mid-17th century to the 1790s, but the outbreak of war with France in 1793 brought a new dimension into English farming which is the subject of the second part of this chapter. The new dimension was government intervention. Until the 1790s war took place on a scale which was not perceived to damage England's island position in resource terms. Relatively little grain was imported - in fact there was an export surplus for most of the period from the 1660s-1790s - and any threats from Continental Europe could largely have been counterbalanced by imports of grain from Ireland (Barnes, 1930). The government had no need to intervene since famine was virtually unknown, supply was adequate to keep prices relatively stable, and even 16th century regulatory measures, such as the Book of Orders, were allowed to fall into abeyance.

The wars with Revolutionary and Napoleonic France between 1793 and 1815 posed the first real threat to supply. In part, this was because the rapid increase in population from the 1770s had turned the grain export surplus into a deficit. By 1790 this was partially compensated by bringing supplies from Ireland, but also, and crucially, by imports from Continental Europe. Napoleon's threat to those supplies forced the government to intervene in agriculture, and brought the first great war time plough up campaign. In 1803 Sir John Sinclair, a Scottish MP and President of the Board of Agriculture, called for an attack on the commons and wastes, which had traditionally been stocked by farmers, but which now seemed to represent an underused resource. 'Let us not be satisfied with the liberation of Egypt, or the subjugation of Malta, but let us subdue Finchley Common; let us conquer Hounslow Heath, let us compel Epping Forest to submit to the yoke of improvement' (Williams, 1970: 57).

Sinclair held a position of influence, his ideas amounted to encouraging government intervention by default. It was a short-term solution, given the expectation that common and waste was unlikely to represent good quality soil, but the message was clear - the ecological balance might need to be compromised for the wider ethical purposes, not in this case of the community, but of government and the nation state. The possibility of compromising sustainability was obvious: Sinclair and his supporters would not have used our terminology, but they would certainly have known they had no way of being sure what the long-term impact would be of wasteland enclosure.

The government and agriculture after 1815

Government intervention was to a short term, narrowly focussed end, but since 1815 governments have been unable to detach themselves from agriculture. In the post-war years duties on grain imports were retained, despite the prevailing atmosphere of free trade, largely to protect farmers or, perhaps more accurately, their landlords. But it ignored the plight of those small owners who had responded patriotically to the call to plough up everything in sight, and were now faced with wartime costs - especially interest rates - in peacetime conditions. More importantly, intervention meant all farmers were operating in the context of government policy, rather than according to the dictates of local conditions. The government finally abandoned the much detested (in the towns) Corn law in 1846 in the wake of the Irish famine, but by then the damage had been done, damage hidden by the relative prosperity of farming in the mid-Victorian ‘golden age’ decades (Collins, 2000).

Having by now seen what it perceived to be the virtues of free trade, Westminster became convinced that future British prosperity lay with industrial and commercial success underpinned by cheap food. Consequently, it chose not to intervene when the golden age gave way to depression brought on by unrelenting and unregulated competition from north and south American grain and meat imports (Turner, 2004: 134-138). The impact on land prices, rents and population, notably the ‘flight from the land’, is well known, and reflected the impact on farmers who were having to deal with external forces (imports) over which they had no control and to which they could not adapt.

Farming was effectively abandoned by the government until the outbreak of the First World War in 1914. Initially imperial preference seemed likely to see the country through, but by 1916 the policy was changing towards plough up and restriction. Again, the government tried to take control but without any real idea of the likely outcome in sustainability terms. The need to supply food to a nation engaged in fighting a war meant sustainability might have to be compromised in the short term. After 1918 the Wheat Act appeared to give farmers protection in the transition from war to peace time, but the government reneged on this support in 1921, and the inter-war years saw depression among the farming community only occasionally countered by government intervention such as the milk, egg and potato marketing boards.

Intervention was again necessary in the Second World War, and this time there was no going back. The 1947 Agriculture Act in effect saw the government taking over agriculture. Expressed loosely, since 1947 a combination of the government (through the Ministry of Agriculture, subsequently Agriculture, Fisheries and Food, and now Defra) and the European Union, have left as little as possible to the individual farmer. Today the farmer does not take his or her own decisions, s/he lives within a context of initiatives and subsidies, plough up and grass down, set aside and the Common Agricultural Policy (CAP).

S/he is vilified for being subsidized, and blamed for all the ills of BSE, CJD, foot and mouth, and a host of other things, which provide a context in which a clear and non-emotional debate about genetic modification seems to be impossible (Martin, 1999).

Government intervention and sustainability

The external economic conditions impacting on farmers in the 19th and 20th centuries were different from those of the 17th and 18th centuries. Yet there is a sense in which the evidence relating to open field England raises pertinent questions about intervention and policy, particularly in relation to the capacity and ability of farmers to control their own environment.

Farmers have always understood the concept of sustainability, even if they have not used the same language that we employ today. They worked the land in a husbandlike - to use a widely employed contemporary term - responsible manner for the well-being of their community and land. This implied caring for the land to ensure that the soil would retain its fertility. Consequently it involved farming in a manner agreed to be the best practice in the locality. Although this might have to be enforced, through lease covenants for example, the reward was a share in both the ownership and the management of the resources - equity. Disputes over equitable access to resources were the most common ways in which sustainable agriculture was likely to be compromised, hence the elaborate system of manorial and other local courts to act as regulators and police within the community.

Farmers recognized that any alterations in the farming cycle had potential implications for the ecological balance and the biodiversity of the soil. They would not have used such language but they clearly understood what it meant, because their bottom line was to pay the rent, make a living, and maintain the farm in such a way as to pass it on to future generations. They acted empirically in the absence of theory.

Farmers may not have used the same terminology but they understood that farming had to meet the needs of any particular generation without compromising the ability of future rural generations to meet their own needs. When they looked to make changes they did so cautiously, ensuring at each stage that the economic demands on farming did not compromise the ecological and ethical considerations too much. Only under the stress of war were such threats largely ignored. They would have understood the Brundtland definition of sustainability: ‘Sustainable development is that development that meets the needs of the present generation without compromising the ability of future generations to meet their own needs’ (WCED, 1987: 144).

Farmers had to act in the past within an economic context they neither understood nor could predict. Their only means of acting was empirical, in response to price mechanisms. Once they were forced by prices to take action they did so with great care, wanting clearly to prevent any upset to the ecological balance, and to the equitable distribution of resources within the community. Almost invariably their long term goal was the future of the farm into the next generation, and in parts of the country this was effectively guaranteed through copyhold and similar tenancy agreements. As a consequence, they had every incentive to farm in a husbandlike manner, and for the most part it worked. Considering that England was the first industrial nation, that its population and towns grew faster than anywhere in Continental Europe, and that this happened without significant famine or similarly problematic conditions, this was no mean achievement (Turner *et al.*, 2001).

Are we in danger of taking too much initiative away from the farmer? Is the self-help of earlier generations being replaced by policy driven directives that take away decision making, in a manner which must damage the capacity of the farmer to take his or her own decisions in the interests of the soil and the farm? If nothing more, we hope that our work will enable policymakers and planners to ask the basic question: if farmers in the past could operate in a manner which we would judge to have been agriculturally sustainable, are we in danger of over regulating? Should farmers in those parts of the world with the greatest problems be encouraged to have more rather than less freedom? What is the right balance between farmer initiative and state intervention?

References

Barnes, D.G. (1930) *A History of the English Corn Laws*. Routledge, London, UK.

Beckett, J.V. (1989) *A History of Laxton; England's Last Open-Field Village*. Basil Blackwell, Oxford, UK.

Collins, E.J.T. (ed.) (2000) *Agrarian History of England and Wales, VII (1850-1914)*. Cambridge University Press, Cambridge, UK.

Conway, G.R. and Pretty, J.N. (1991) *Unwelcome Harvest: Agriculture and Pollution*. Earthscan, London, UK.

ESRC Data Archive, SN 4517. Crop Yields and Animal Carcass Weights in England, c.1700-1914. http://www.data-archive.ac.uk/search/indexSearch.asp?ct=xmlSn&q1=4517

ESRC Data Archive, SN 4537. Ecological Aspects of Agricultural Sustainability, c. 1576-1946: the Archives. http://www.data-archive.ac.uk/search/indexSearch.asp?ct=xmlSn&q1=4537

Martin, J..M. (1999) *The Development of Modern Agriculture: British Farming Since 1931*. Macmillan, Basingstoke, UK.

Sen, A. (1981) *Poverty and Famines*. Oxford University Press, Oxford, UK.

Sen, A. (1986) Food, economics and entitlement. *Lloyds Bank Review* 160(2) 1-20.

Thirsk, J. (ed.) (1984-5) *Agrarian History of England and Wales, V, 1640-1750*. Cambridge University Press, Cambridge, UK.

Turner, M.E., Beckett, J.V. and Afton, B. (2001) *Farm Production in England 1700-1914*. Oxford University Press, Oxford, UK.

Turner, M.E., Beckett, J.V. and Afton, B. (2003) Agricultural Sustainability and Open-Field Farming in England, c. 1650-1830. *International Journal of Sustainable Agriculture* 1, 124-140.

Turner, M.E. (2004) Agriculture, 1860-1914. In: Floud, R. and Johnson, P. (eds) *The Cambridge Economic History of Modern Britain: Vol II, Economic Maturity, 1860-1939*. Cambridge University Press, Cambridge, UK, 133-60.

Williams, M. (1970) The enclosure and reclamation of wasteland in England and Wales in the eighteenth and nineteenth centuries. *Transactions of the Institute of British Geographers* 51, 55-69.

Williamson, T. (2002) *The Transformation of Rural England: Farming and the Landscape, 1700-1870*. University of Exeter Press, Exeter, UK.

WCED (1987) (Brundtland Commission) *Our Common Future*. Oxford University Press, Oxford, UK.

Young, A. (1771) *The Farmer's Tour Through the East of England,* vol. 1. W. Strahan, London, UK.

Chapter 5

Soils and sustainability: the future of the South Downs

J. Boardman

Introduction

The proposed establishment of a new National Park on the South Downs brings into sharp focus questions about competing uses of rural areas and related questions of sustainability. All government departments and agencies have stated commitments to sustainable development. It is also reasonable to suppose that sustainable development will rank high in National Park objectives, with the Council for National Parks (2003: 3) declaring 'National Parks should act as beacons for sustainable development'. Any discussion of 'sustainable use', however, has to recognize the problem of definition. In many official documents 'sustainable development' is used ubiquitously and loosely to mean any desirable end, e.g. a reduction in bullying in the workplace (Defra, 2000). Other documents find it easier to define '*un*sustainable management' especially with reference to soils, e.g. EEA (2000). In this chapter I define sustainable use or management with reference to maintenance of the necessary soil functions for agricultural production, avoidance of off-farm damage, and adequate support for incomes dependent on utilization of the soil. Sustainability has a time dimension and also a spatial context, which I shall discuss.

In the national context there is concern about the costs of damage to natural resources from agriculture, estimated at £1.2 billion annually (EA, 2002), and the sustainable use of soils. Estimates suggest that 18% of soil organic carbon in arable topsoils has been lost between 1980 and 1995; also, costs of erosion are estimated at £23 to £50 million per annum (Defra, 2004: 7).

Pressure on land use is not a new issue on the South Downs, for instance the well known bungalow development of Peacehaven after the First World War (Brandon, 1999: 171) and expansion of urban areas onto the Downs in the inter-war years (with implications for the soils-erosion-flooding story).

The South Downs is a special case in terms of National Park designation. It is easily accessible from London, it contains long-established holiday resorts, such as Brighton and Worthing, and therefore a large population partly dependent on income from a recently estimated 39 million day visits per year (South Downs Focus, 2005). Popular perception of chalk downland landscape is an important consideration which has motivated calls for National Park status. The view of extensive areas of rolling chalk grassland grazed by sheep and rabbits is a throw-back to inter-war times when arable agriculture was depressed. At various times in the past the South Downs was dominated by sheep grazing but at others it was a cereal growing landscape. Many people now have a view of what this semi-natural landscape should look like based on a past that has not existed for over 50 years. However, others acknowledge these issues but still argue that in modern Britain a South Downs dominated by pasture is desirable from several points of view. The implications of a sustainably managed chalk downland landscape will be explored in this chapter.

What criteria have to be satisfied for us to accept that soils are being used sustainably? First, that the soil is not damaged to such an extent as to make it unusable by future generations. In this regard we have to assume that they will want to use the soils in similar ways as at present: arable farming, grazing etc. This is thus a matter of not significantly compromising their choices as to future use - but choices are already compromised on the South Downs by past usage. Second, that those dependent on the soil and the landscape can continue to make a living from its use - or that viable alternatives are provided e.g. tourist income for farming communities. Third, that the multifunctionality of the soil and the landscape is recognized and taken into account. Thus while farming may be a major use, other uses are recognized and encouraged; the 'Right to Roam' legislation is an example. However, multifunctional thinking implies constraints on particular uses so that they do not compromise other potential uses; for example, that farming should not pollute the aquifer by over-use of nitrates and pesticides. Fourth, that the off-site impacts of farming do not have unacceptable costs for local communities, local authorities, water companies, nature reserves, the taxpayer, etc.

Erosion history, biodiversity and the sustainable use of soils

At the beginning of the Holocene, the chalk downlands of southern England were covered by a considerable thickness of loess and therefore, by implication, by deep, fertile, easily-eroded soils (Catt, 1978). Eroded sediments in valley bottoms and on flood plains, on coastal plains such as Selsey, occasional buried soils and comparisons with other areas, support this contention. Catt (1978) estimated a loess cover of up to 4 m; Favis-Mortlock *et al.* (1997) obtained reasonable results in a long-term modelling exercise by assuming initial thicknesses of 1m. The loess cover sustained deciduous forests (Thorley, 1981; Waller and Hamilton, 2000) and their removal by humans and replacement by grazing and arable uses, precipitated erosion that has continued intermittently for the last 6000 years. Bronze and Iron Age communities seem to have been responsible for major clearance and consequent erosion (Favis-Mortlock *et al.*, 1997). Erosion was driven by land use, not principally climate, and occurred at different times in different places (Bell, 1982; Evans, 1996: 74-79). Thus it is ironic that the grassland for which the Downs became famous is a result of erosion and the loss of depth and quality of the soils which, until intensive post-war farming practices were introduced, were not suitable for arable crops.

The removal of the natural vegetative cover from much of the South Downs led to replacement by a changing mosaic of grassland, woodland, arable and settlements. Ironically, in recent times, the biodiversity value of chalk grassland has attracted great attention even though it is recognized as a plagio-climax community maintained artificially by sheep and rabbit grazing. However, its scarcity internationally and its dramatic losses in England have added to its perceived value. Burnside *et al.* (2003) report that there are currently about 2900 ha of chalk grassland covering about 3% of the Sussex Downs. Much of the grassland is now fragmented and confined to marginal land and scarp slopes. Other areas of grassland are devalued (species-poor) through addition of fertilizers.

It is however true that the perceived value of the Downs is often stated in terms of a cultural landscape with archaeological, artistic and literary connotations (Brandon, 1999). We maintain, and are currently seeking to extend, protection for an area that is high in cultural, recreational and landscape value, lacks natural vegetation, but has grassland-associated communities with international rarity value. These points should be kept in mind when considering the sustainable use of soils because they raise the issue of what is the particular character of the soils that we wish to protect and for what future use.

Future use of soils

Future use is a difficult issue but is core to the problem of sustainability. It is easiest to assume that present trends will continue but that National Park designation, if it occurs, will tend to encourage recreational use of the landscape and will discourage arable farming uses. These trends are already in place with support for non-arable conversions through the Set Aside mechanism, the Environmentally Sensitive Area (ESA) funding and the new Environmental Stewardship scheme. A powerful present trend is the decline in cereal prices so that cultivation of marginal land, such as much of the South Downs, is less economic than previously. To put this in context, average winter wheat yields on the Downs are around 7 t/ha with nitrate inputs and this compares unfavourably with yields on better soils in eastern England which reach 10 t/ha or more. It is therefore arguable, that in a national context, we should not use the South Downs for arable farming but should find alternative ways to support farmer income, such as sheep grazing, conservation and recreational uses. Mixed farming, with a careful selection of sites for arable usage, may be a viable and desirable option but at the present time the Downs are losing medium-sized farms (SDCB, 2005: 35). I shall return to this issue.

Damage to soils

The question of damage to soils is central to the issue of sustainability. Soil erosion causes damage because it reduces the thickness of the soil and therefore its water holding capacity and its store of plant nutrients. The A horizon with high organic content is particularly important but is the first to be eroded. The impact of the damage that erosion causes will vary depending on soil thickness. The South Downs soils are particularly vulnerable because they are typically less than 25 cm thick, already stony, and overlie chalk. On thicker soils, such as those on the Lower Greensand, there is less immediate threat to productivity. Continuing erosion of chalk soils means that they become stonier, more droughty and less easy to work. This does not pose an immediate problem because of current low rates of erosion (see below), but if sustainability is interpreted to include questions of the long-term viability of farming, then this is relevant.

Rates of soil formation, against which we should compare losses, seem to be very low, especially under arable farming systems where little organic matter is returned to the soil after crop harvesting. In essence, the soil is a non-renewable resource in the time-scale of human usage (e.g. UNEP, 2004: 7). The chalkland soils will be renewed by deposition of loess in the next glacial period. In the meantime continued arable farming (with ploughing) increases the stoniness of these shallow soils and reduces organic content. For example, a 400-year EPIC model simulation predicted an increase in stone content from 38% (present average) to 86% (Favis-Mortlock *et al.*, 1991). Comparisons of organic carbon levels in permanent grassland near Lewes, East Sussex, with arable fields, show a decline from 11.9% to 3.1% over a 40 year period, due to mixing with mineral material, oxidation and erosion (Boardman and Robinson, 1985). As soils become more stony, erosion rates will decrease due to an increase in infiltration but soil fertility and the efficacy of fertilizers will also decrease. Farming costs will increase and arable farming may not be viable. With soil formation rates effectively zero the question arises as to whether current erosion rates are a threat to sustainability (here narrowly defined as the profitable continuation of arable farming) and whether other systems of farming can minimize erosion.

Current soil erosion rates on the South Downs

An area of about 36 km^2 of farmed land was monitored from 1982-91, with all instances of erosion being recorded (Boardman, 2003). The period includes two relatively wet autumns, 1982 and 1987, when there was extensive erosion on winter cereal fields. Since 1991, generally dry autumns and less cultivation of winter cereals has led to less erosion with the exception of 2000-01 (Boardman, 2001). Erosion rates on eroding arable fields vary from 0.5 to 5.0 m^3/ha/year. Such rates do not, in the coming decades, threaten the ability of the farmer to produce arable crops such as wheat, oil seed rape and linseed. Only occasionally, as a result of exceptional runoff, does erosion impact on farming operations with the formation of deep rills and ephemeral gullies. These may interfere with fertilizer applications and with harvesting. For example, at Rottingdean in 1987 rates of over 200 m^3/ha were recorded on one 12 ha field (Boardman, 1988a).

The issue of sustainability (can we grow crops on poor soils?) then becomes an economic question: does society wish to support this activity financially? Unfortunately the farm is not a closed system and impacts, both immediate and longer-term, are felt elsewhere. The sustainability of enterprises dependent on the soil has therefore to be considered in a wider spatial context (Boardman *et al.*, 2003).

Off-farms impacts

The South Downs in common with other calcareous terrains has little overland flow in the form of permanent watercourses. The exceptions are major rivers: Arun, Adur, Ouse and Cuckmere which cut through the Downs in broad valleys, and ephemeral streams ('winterbournes') such as the Lavant and the Lewes Winterbourne whose systems were more extensive before the onset of Victorian pumping reduced the level of the water table in the valleys. The Downs, therefore, does not have the threat to watercourses of pollution by sediment and agricultural and urban chemicals causing eutrophication of lakes and rivers. Nor are there dams and reservoirs to be sedimented. In other parts of southern England sediment pollution is a major threat to freshwater fish stocks, e.g. in the River Rother in West Sussex. There is also a limited range of crops grown on the Downs due to the thin, infertile soils. Crops such as maize, potatoes and sugar beet are not viable and therefore erosion and pollution associated elsewhere with these crops is not present here.

Thousands of years of erosion means that the thin, stony, downland soils require fertilizers and farmers require adequate prices to continue their operations. Nitrates pollute water supplies and levels in drinking water have to be reduced below EU threshold standards. Pretty *et al.* (2000) estimate that nitrate in drinking water was costing the UK £16 million a year in 1996. Groundwater carrying nitrates travels through the chalk at about 1 m per year to a saturated zone which may be 50 m below the surface (Jones and Robins, 1999). High levels of nitrate application associated with the post-war years may not yet have reached the water table; there is thus a considerable time lag in these effects (Foster and Grey, 1997). At the present time, of nine boreholes monitored for nitrate levels on the South Downs, eight have an upward trend. The Water Framework Directive requires a level or downward trend by 2015 (SDCB, 2005: 25).

The most serious impacts of erosion and runoff from agricultural land are the muddy floods that have regularly affected urban areas in the last 30 years. Muddy floods are recognized in many European countries as a consequence of intensive agriculture and the cultivation of erosion-prone soils for certain crops (Boardman *et al.*, 2006). On the South Downs, the threat was first recognized in the early 1980s (Stammers and Boardman, 1984). The main cause is the growing of winter cereals and hence bare smooth surfaces at the wettest time of the year. Damage to property occurs because of the close proximity of urban areas and arable land. Damage is concentrated in the wetter autumns with 14 incidents of flooding in 1982, 34 in 1987 and 53 in 2000 (Boardman, 2003, Fig. 1). The scale of the damage varies from the inconvenience of flooded roads to the inundation of groups of houses. The worst case was the flooding of Rottingdean in 1987 with 74 houses affected and estimates of expenditure by householders of £300,000 and £153,000 by the local authority, excluding fire and police service costs. Of the costs to individual householders, £112,000 of uninsured losses had to be borne by them (Robinson and Blackman, 1990). Areas such as Sompting have been affected by repeated flooding from the same source (Evans and Boardman, 2003). Breaky Bottom vineyard and farmhouse was flooded in 1972, 1976, 1982, 1987 and 2000-01, in the latter year on 21 occasions (Boardman, 2001). The 1987 and 2000-01 damage was the subject of an out-of-court settlement. It is clear that the use of soils in such a way as leads to regular muddy flooding affecting urban areas violates a principle of sustainability. Similarly, if the building of new housing estates led to runoff causing erosion and flooding of farmed downslope areas, farmers would be right to claim recompense.

The sustainable use of soils on the South Downs

Three distinct but related issues of the sustainable use of soils have been discussed in this short review. The first is the loss of chalk grassland, a valued but declining biological resource. The answer here is protection of remaining grassland and appropriate management strategies such as selective grazing. There are also issues of fragmentation and therefore distance between remaining reserves (Burnside *et al.*, 2003). This issue interacts with problems concerning the impact of arable farming on water and urban communities since the loss of chalk grassland over several decades has contributed to the 'arablization' of the landscape. The reverse process can only help solve some of these problems. The second issue is nitrate pollution of groundwater. Advice to farmers in recent years has been to avoid spreading nitrates on cereal crops in the autumn when uptake is low. Decreases in the area of cereals in the future will lead to less pollution as will extension of the current Nitrate Vulnerable Zone on the Downs (SDCB, 2005). Reduction in nitrate application will benefit the restoration of chalk grassland in that nitrates encourage some species to out-compete others (competitive exclusion). Third, the threat of muddy floods can be reduced by selective return to grass of fields in valley bottoms and on steep slopes. Evans and Boardman (2003) show that a relatively small return to grass in the Sompting catchment (15%) greatly reduced the risk of flooding. This was because patterns of valley bottom flow ('connectivity') were interrupted. Such targeted interventions are attractive both because they are effective in an off-farm sense and because they have minimal impact on farm incomes; for example, set aside has a minimal impact because it is funded by a government subsidy. They also reduce, in the case of Sompting, the inevitable recourse to the law that further flooding would have led to. But we have been here before: 'some form of targeting of grant monies on high-risk sites is desirable' (Boardman, 1988b: 44).

Other solutions to these problems have been proposed and are worthy of mention. Erosion and runoff are greatly encouraged by bare soil surfaces. An approach using minimum tillage where soils are not inverted by the plough is therefore attractive. Unfortunately there is some evidence that while erosion may be reduced, runoff may actually increase due to more compact soils. Also, herbicides to control weeds in minimum tillage systems may result in groundwater pollution (Evans, 2005). Such schemes need extensive trials, which include wet autumns.

Major flooding and erosion problems at Rottingdean, Sompting and Breaky Bottom, have in common significant gullying, due to large volumes of runoff from catchments dominated by winter cereals. Thus, extensive areas were bare of vegetation cover in the critical autumn months and systems of rills and gullies were established on them. We do not know what the effect would have been of minimum tillage throughout the catchments. However, we do know that selective change to grass in critical areas reduces erosion, helps build better, structured topsoils, does not impact significantly on farmer income and also offers protection from costly legal action. A move to a more mixed farming economy in areas dominated by arable land use offers several advantages and in future is likely to attract government funding through agri-environment schemes.

Discussion

The interim management plan for the South Downs adopts the following definition of sustainable development: 'a dynamic process that enables all people to realize their potentials and to improve their quality of life in ways that simultaneously protect and enhance the Earth's life support systems' (SDCB, 2005: 4). This is woolly, idealistic and unrealistic in a world of conflicting land uses and soil functions. The interim plan has very little to say about soils. The Soil Action Plan for England recognizes that, 'often different soil functions are not mutually compatible' (Defra, 2004: 9).

'Sustainable use' implies consensus or at least democratically arrived-at decisions as to what is a sustainable use! This raises the issue of which constituency we are addressing. Even the issue of the desirability of a National Park (let alone sustainable use) is contested: thus the need for the current Public Inquiry. A National Park and more sustainable policies come at a price and would shift the balance of support to national rather than local funding bodies. As the Interim Plan remarks: 'there is no shared vision for the South Downs with wide public backing' (SDCB, 2005: 60).

The sustainable use of soils cannot be achieved by addressing the problem on a field by field basis. The unit of decision making is the farm, and a farmer is likely to have a plan for the whole farm rather than for individual fields. What we can hope is that farmers will take into account erosion and off-farm problems when planning land use. MAFF (1999) suggested that farmers identify fields at risk on the basis of simple risk factors such as soil type and slope. However, flooding of communities is likely to be an issue at a catchment scale requiring the cooperation of several farmers, householders and local authorities (Evans and Boardman, 2003). Sadly there is no mechanism that encourages this process and success seems to be as much a function of the threat of legal action and the current price of cereals as a desire to act responsibly towards neighbours. We await with interest the impact of the newly created Catchment Sensitive Farming scheme for 40 problem catchments in England and Wales and the results of the small number of Landcare partnerships (e.g. Defra, 2006). The English tradition of a lack of planning control over agricultural land use, be it benign or damaging, is unhelpful in terms of planning for sustainability. The Soil Action Plan ignores this issue, regarding planning control of land use as a means of protecting soils from urban developments, and control of unsuitable agricultural use as a matter for better farmer advice and financial incentives in a reformed CAP (Defra, 2004). Current changes in agricultural funding mean that under the Single Payment farmers will have to keep their land in 'good agricultural and environmental condition'; they will have to be aware of the *Guidance for Soil Management*, and will have to devise a Soil Protection Review. If farmers apply for the Environmental Stewardship scheme, they will have to complete a soil risk assessment as part of their Farm Environment Record (Defra, 2005).

At the level of local authorities, solutions to the problems of muddy flooding are greatly hampered by a lack of institutional memory and geographical awareness. Local Authority officers act independently and without knowledge of similar happenings in adjacent areas. There is little exchange of information or transfer of experience, thus in the event of flooding a new learning curve has to be created. In a national context, a move to sustainable management in specified areas may impact on other areas where intensification and pollution could increase. However, with the developing EU legislative framework for nitrates, water quality and plans for a soil strategy, this may be an unlikely scenario.

Conclusion

In the inter-war years, decisions regarding the expansion of urban areas onto the Downs were subject to debate as regards the aesthetics of houses on the skyline. Most expansion occurred along dry valleys and on valley-side slopes. At the time there was no threat of flooding from farmland because most of it was under grass. With post-war moves into arable agriculture and, in the 1970s into winter cereals, these trends now appear to have been unfortunate. The loss of chalk grassland and nitrate pollution were driven by the intensification of farming and maximization of cereal production. More recent developments such as the fall in the price of cereals, revisions to the CAP, greater awareness of off-farms impacts, damage to archaeological sites, and the growing clamour for a National Park, have combined to offer an opportunity for the arguments for sustainable management of soils and landscape to be heard. The challenge remains to find the mechanisms to put these ideals into practice. Many of the problems would be solved by the encouragement of a mixed farming economy with targeted grant aid for protection and enhancement of chalk grassland and consequent reduction of the risk of erosion and muddy flooding. The benefits of such an approach would be to the farming community and wider society. Measures on soil erosion, soil organic matter content and soil structure will be among the cross compliance conditions attached to the Single Payment under the reformed

CAP (Defra, 2004: 10) and soil management taking into account sustainability issues is central to the proposed EU Directive establishing a framework for the protection of the soil (EC Europa, 2006).

Acknowledgements

I thank Drs Bob Evans and Niall Burnside and Professor Mike Edmunds for helpful comments on a draft of this chapter.

References

Bell, M. (1982) The effects of land-use and climate on valley sedimentation. In: Harding, A.F. (ed) *Climatic Change in Later Prehistory*. Edinburgh University Press, Edinburgh, UK, 127-142.

Boardman, J. (1988a) Severe erosion on agricultural land in East Sussex, UK, October 1987. *Soil Technology* 1, 333-348.

Boardman, J. (1988b) Public policy and soil erosion in Britain. In: Hooke, J.M. (ed.) *Geomorphology in Environmental Planning*. Wiley, Chichester, UK, 33-49.

Boardman, J. (2001) Storms, floods and soil erosion on the South Downs, East Sussex, autumn and winter 2000-01. *Geography* 86(4) 346-355.

Boardman, J. (2003) Soil erosion and flooding on the eastern South Downs, southern England, 1976-2001. *Transactions Institute British Geographers* 28(2) 176-196.

Boardman, J., Poesen, J. and Evans, R. (2003) Socio-economic factors in soil erosion and conservation. *Environmental Science and Policy* 6(1) 1-6.

Boardman, J. and Robinson, D.A. (1985) Soil erosion, climatic vagary and agricultural change on the Downs around Lewes and Brighton, autumn 1982. *Applied Geography* 5, 243-258.

Boardman, J., Verstraeten, G. and Bielders, C. (2006) Muddy floods. In: Boardman, J. and Poesen, J. (eds) *Soil Erosion in Europe*. Wiley, Chichester, UK, 743-758.

Brandon, P. (1999) *The South Downs*. Phillimore, Chichester, UK.

Burnside, N.G., Smith, R.F. and Waite, S. (2003) Recent historical land use change on the South Downs, United Kingdom. *Environmental Conservation* 30(1) 52-60.

Catt, J.A. (1978) The contribution of loess to soils in lowland Britain. In: Limbrey, S. and Evans, J.G. (eds) *The Effect of Man on the Landscape: the Lowland Zone*. CBA Research Report No. 21, London, UK, 12-20.

CNP (2003) *Strategic Plan 2003-2008*. Council for National Parks, London, UK.

Defra (2000) *Sustainable Development Action Plan: Just Jump Straight In*. Department for Environment, Food and Rural Affairs, London, UK.

Defra (2004) *The First Soil Action Plan for England: 2004-2006*. Department for Environment, Food and Rural Affairs, London, UK.

Defra (2005) *Entry Level Stewardship Handbook*. Department for Environment, Food and Rural Affairs, London, UK.

Defra (2006) *Case Studies aimed at Reducing Diffuse Water Pollution from Agriculture in England*. Department for Environment, Food and Rural Affairs, Water Quality Division, London, UK.

EA (2002) *Agriculture and Natural Resources: Benefits, Costs and Potential Solutions*. Environment Agency, London, UK.

EC Europa (2006) *A Strategy to keep Europe's Soils Robust and Healthy*. [Accessed Dec 2006] Available from http://ec.europa.eu/environment/soil/index.htm

EEA (2000) *Down to Earth: Soil Degradation and Sustainable Development in Europe*. Environmental Issue Series No 16, European Communities, Luxembourg.

Evans, R. (1996) *Soil Erosion and Its Impacts in England and Wales*. Friends of the Earth, London, UK.

Evans, R. (2005) Reducing soil erosion and the loss of soil fertility for environmentally-sustainable agricultural cropping and livestock production systems. *Annals of Applied Biology* 146, 137-146.

Evans, R. and Boardman, J. (2003) Curtailment of muddy floods in the Sompting catchment, South Downs, West Sussex, southern England. *Soil Use and Management* 19, 223-231.

Favis-Mortlock, D.T., Boardman, J. and Bell, M. (1997) Modelling long-term anthropogenic erosion of a loess cover: South Downs, UK. *The Holocene* 7, 79-89.

Favis-Mortlock, D.T., Evans, R., Boardman, J. and Harris, T.M. (1991) Climatic change, winter wheat yields and soil erosion on the English South Downs. *Agricultural Systems* 37, 415-433.

Foster, S.S.D. and Grey, D.R.C. (1997). Groundwater resources: balancing perspectives on key issues affecting water supply and demand. *Journal C.I.W.E.M.* 11, 193-199.

Jones, H.K. and Robins, N.S. (eds) (1999) *The Chalk Aquifer of the South Downs*. Hydrogeological Report Series of the British Geological Survey, UK.

MAFF (1999) *Controlling Soil Erosion: A Manual for the Assessment of Agricultural Land at Risk of Water Erosion in Lowland England*, Ministry of Agriculture, Fisheries and Food, London, UK.

Pretty, J.N., Brett, C., Gee, D., Hine, R.E., Mason, C.F., Morison, J.I.L., Raven, H., Rayment, M.D. and van der Bijl, G. (2000) An assessment of the total external costs of UK agriculture. *Agricultural Systems* 65, 113-136.

Robinson, D.A. and Blackman, J. D. (1990) Some costs and consequences of erosion and flooding around Brighton and Hove, autumn 1987. In: Boardman, J., Foster, I.D.L. and Dearing, J.A. (eds) *Soil Erosion on Agricultural Land*. Wiley, Chichester, UK, 369-382.

SDCB (2005) *Interim South Downs Management Plan*, co-ordinated by the South Downs Conservation Board. Arundel, West Sussex, UK.

South Downs Focus (2005) 14 January 2005. East Hampshire AONB Project/Sussex Downs Conservation Board.

Stammers, R. and Boardman, J. (1984) Soil erosion and flooding on downland areas. *The Surveyor* 164, 8-11.

Thorley, A. (1981) Pollen analytical evidence relating to the vegetational history of the chalk. *Journal of Biogeography* 8, 93-106.

UNEP (2004) *UNEP's Strategy on Land Use Management and Soil Conservation*. UNEP, Nairobi, Kenya. [Accessed Dec 2006] Available from http://www.unep.org/pdf/UNEP-strategy-land-soil-03-2004.pdf

Waller, M.P. and Hamilton, S. (2000) Vegetation history of the English chalklands: mid-Holocene pollen sequence from Caburn, East Sussex. *Journal Quaternary Science* 15, 253-272.

Chapter 6

The role of organics within sustainable farmland management

M. Reed

Introduction

During the 1990s a new tool was added to the locker available for sustainable land management - the shopping basket. As a means of influencing the way in which farmland is managed it is an unlikely but increasingly powerful device. That a group of consumers came to understand that their shopping choices have a direct relation to the quality of agro-ecosystems is the central innovation of contemporary organic farming. From the early 1980s the organic movement firmly linked the success of organic farming to the expansion of the market for their goods, and in doing so dramatically widened the scope of those with a stake in the sustainable management of farmland. Marrying that marketing with the scientific evidence that underpins managing farmland in a manner that sustains or improves its agro-ecology has proved to be a significant challenge. The organic movement also introduced an element of politicization to the management of food and land use, mixing marketing and campaigning in a way that has disconcerted many observers.

The dynamism and speed of the rise of the British organic movement has meant that it is difficult sometimes to assess its significance. In no small part this is because it is often enmeshed in the controversies of the moment, with proponents and detractors equally passionate about it, making an assessment difficult. This chapter brings a range of social scientific perspectives to bear on the rise of the British organic movement that are not often used in relation to land management. It considers three elements of organic farming that have made it so distinctive and the interactions between them. Before proceeding it is important to note that the newness of this phenomenon means that many social scientists, like farmers and land managers, are having to adjust to innovations being created in the fields of organic food and farming. Life always runs in advance of theory.

Organic farming as it is currently configured is a dynamic combination of three complex elements - a social movement, a brand and a regulatory system. A social movement is a loose collective of people bound together by a set of views of the world and the willingness to take action together in pursuit of them. The second element is that this group of people constructed a form of 'brand' around the produce from their farms, holdings, restaurants and kitchens. This brand, they hope, exemplifies and embodies what they believe, allowing the end goals of the movement to be realized in the present, but it also allows people to take part at a distance. The final and most recent pillar is the regulatory system that underpins the brand and to a degree validates some of the claims made by the movement.

Many accounts of the rise of the English organic sector point to either the huge rise in the retail sales of organic foods, from £1 million in 1985 to £1.2 billion in 2005, or to the area of land under organic management, 686,101 ha in 2005 compared with less than 50,000 ha in 1997, or the number of holdings, 4010 in 2005 compared to 828 in 1997 (Lobley *et al.*, 2005). Impressive though these figures are as a suggestion of the rapid growth of the sector, they do not reveal the involvement of the consumer in this process. In most polls 75% of consumers say that they buy organic produce; in more recent findings 48% of shoppers in occupational class E say they 'eat organic' (Soil Association, 2003). In 2003, 81% of the retail sales of organic food were through the supermarkets but by 2004 this had fallen to 75%, with rapid growth in box schemes, farm shops and farmers' markets (Soil Association, 2005). However, most organic consumption is concentrated in a small group of shoppers who buy most organic produce; this group represents around 20% of shoppers but they buy about 80% of organic produce. If we extrapolate these figures, they suggest that several million people are involved in buying organic food and to some degree in supporting organic production.

The success of the organic movement in the last 20 years has been to recruit a large group of dedicated consumers who are prepared to pay more for what they see as the benefits of organic food - amongst which is an improved environment. In most accounts of the food chain, the end consumer is viewed as a passive recipient, somewhat flighty in the face of trends and fads. This view of consumers is often shared by many social scientists who view consumption as the antithesis of living a whole and real life (Miller, 2001). In this chapter I am going to suggest an alternative view, that consumption can be an act in which we not only create our identities but can express a certain degree of community (Valentine, 1999). That consumption links to a new form of politics, in the sense of how we lead our lives, and that organic food is part of this change. This means that the management of farmland is something that millions of people are concerned with and that this is mediated through the consumption of organic food.

The standards

> Foodstuffs will have to be graded, marketed, and retailed according to the way the soil is manured (Howard, 1940: 221)

Albert Howard, one of the founders of the early organic movement looked forward to a time when the method of production would determine how agricultural produce would be marketed (Conford, 2001; Moore-Colyer, 2004). For many years, however, within the organic movement most of its efforts were placed on trying to prove the scientific superiority of organic food so, it hoped, making a marketing strategy unnecessary. By the late 1950s the early retailers of organic foods started to create their own standards on behalf of their customers, and this was matched by a proto-type standards scheme in the United States. In the UK the first recognizable standards for how organic production should be conducted were instituted in 1967 and took the form of an agreement that the farmer or grower signed, agreeing not to use a proscribed range of chemicals in return for being listed in a register of producers. This 'declaration of intent' was really a gentleman's agreement and at the time did not seem to gain much purchase within the movement. The foundation of the contemporary standards came with the intervention of Fritz Schumacher when he was President of the Soil Association in the early 1970s. Previously Schumacher had worked for the National Coal Board and it had instituted a system by which the quality of coal was improved by certifying coal merchants in return for allowing them to display a symbol indicating that their coal met these standards (Reed, 2004). In his determination to make the Soil Association and the rest of the organic movement more worldly, Schumacher engineered the transfer of his coal scheme to organic production.

As quickly as these standards were introduced some people dissented from them in diverging directions; for some they were hopelessly uncommercial and for others they were a compromise too far. The standards signalled not just what was not permissible in organic farming but implicitly an orientation and strategy as to how organic farming was to grow. The Soil Association's symbol scheme, allowing farmers and growers to display the approved logo, introduced an inspection system, created a brand and signalled the importance of commerce in advancing organic farming. In what was then a small movement - in the 1970s the Soil Association had less than 5000 members - it caused years of discussion and argument.

Much of the debate and dissension within the organic movement was brought to a halt by the intervention of the European Community. In response to a planned directive (EC 2092/91) the then Thatcher government moved to institute domestic law to regulate the sale of organic produce. Schumacher had wanted to use the market to grow the organic movement but his market was that of small businesses and co-operatives, quite different from the neoliberalism of Thatcherism (Dudley, 1991). In forming the UK Register of Organic Food Standards (UKROFS) the attempt to institute a market mechanism into organic food was blatant. UKROFS would provide the baseline standards for organic production, it would also authorize agents who could certify crops or farms as being in compliance with those standards and would oversee the development of the standards over time. The board of UKROFS would comprise civil servants, representatives from the organic movement, academics and representatives of the food industry. False claims about organic production would now be investigated by Trading Standards Officers and arrangements were made to ensure that imports met the required standards. Although the organic movement had representation on the UKROFS board it had no right to representation and it was clearly hoped that, as in the US, commercial certification agencies would appear (Blake, 1994). With the sudden attention of others the organic movement found a discipline it had previously lacked.

Legally enforceable standards of production transformed the organic sector. It provided a creditable set of guidelines for farmers and growers to follow but also for policymakers to engage with and to consider. The increasingly important multiple retailers undermined the Thatcherite goal of a market as they backed the guidelines of the Soil Association, and its smaller rival the Organic Farmers and Growers, as they were simple for their customers and logistics systems to understand. Policymakers could design measures to assess the environmental impacts of organic farming and in time offer rewards for the extra gains that it made. Doubts about provenance from customers could be referred to the strictures of trading standards and the organic organizations benefited from the revenue flows from inspection. The standards system did become a vehicle for promoting and extending the support for organic farming but in way that Schumacher and its other architects could never have envisaged.

Although the organic movement prided itself on its environmental credentials, others were not necessarily convinced. Explicit interest in the environmental impact of organic farming started in the 1960s but this did not travel outside of the organic movement. In part this was because the wider conservation movement was more interested in reserves from farm land rather than the improvement of the agri-environment of the land itself. This led to a public dispute in the 1980s between the Soil Association and the RSPB over the then new policy of set-aside. At this time the first scientific enquiries into how sustainable organic farming actually was were published (Arden-Clarke and Hodges, 1987; 1988). Gradually evidence has accrued about the environmental impacts of organic farming, leading Lampkin (2005: 193) to argue that 'There is now a significant body of research indicating the beneficial effects of the organic practices on soil structure, organic matter levels and biological activity, as well as plant, insect and bird and wild animal biodiversity'. Simultaneously the organic movement has entered an on-going dialogue with conservation bodies to avoid further public disputes. That does not, however, mean that the scientific evidence about the sustainability of organic farming has become uncontentious.

The movement

Organic farming or organic agriculture is often referred to as the organic sector without a great deal of analytical attention being paid to its constituent parts. Within the sector farmers, growers and consumers often differentiate between the industry and the movement, again without much analytical attention. Understanding how the sector is constituted is key to understanding the novelty of its challenge and reasons why it has been able to transcend the normal parameters of British agricultural politics. Broadly we can say that there are two intertwined and interactive components. One is the industry which comprises farmers, growers, processors, retailers and consultants who make most or all of their living from organic farming and food. Aligned with this group is the organic movement, the social movement of consumers, growers, gardeners and activists who ally themselves with the industry. Many of those in the industry are also members of the movement, in that they passionately believe in its aims, and it is this 'dual membership' that lends the industry and the movement its particular disposition.

The general conceptualization of a social movement has altered substantially in the last 20 years, with a synthesis being established between the generally North American focus on the pragmatics of movement activity and the European emphasis on the meaning of it. This new synthesis has found its most sophisticated expression in the work of della Porta and Diani (1999):

- *Informal interaction networks* - interaction between individuals, groups and organizations; a range of networks from loose to dense, as a precondition and setting for:
- *Shared beliefs and solidarity* - the symbolic redefinition of social problems and the formation of a new collective identity.
- *Collective action focusing on conflicts* - promotion/opposition to change, usually around the contestation of a social stake.
- *Use of protest* - unusual political behaviour, not just voting but other forms of frequent protest activity.

Whilst it is not necessary to pursue what characterizes a social movement at great length at this juncture, it is important to note how the features above map onto the organic movement and the mobilization against Genetically Modified (GM) crops provides an excellent example (Reed, 2002). Many of the protests, petitions and meetings were held by impromptu groups, formed by informal networks of friends and acquaintances, which already shared many beliefs and were acting in no small part in solidarity with organic farmers and growers. The collective actions ranged from petitions, boycotts of GM products and the threat of a wider consumer boycott in order to gain some control of the food chain (the social stake) and at the most extreme, protestors were prepared to go to prison to demonstrate their strength of feeling. Many of these aspects of social movement activity were already present but were made public by the needs of attempting to defeat the introduction of GM crops (Reed, 2006).

Without being cognisant of the ability of the organic movement to mobilize actors beyond the normal realms of agricultural politics its rise is hard to explain. Parts of the organic movement are permanently campaigning, looking for issues through which they can push its message forward. This runs alongside the marketing of organic produce to the point that often it appears as if the two are connected. Using the GM example again, much of the movement was deeply engaged in protesting against the new technology and actively used organic as the obvious and existing alternative. At the same time the packaging of organic goods, and many shop posters, emphasised their 'GM free' status and urged consumers to become involved in protesting against the crop trials. In this way marketing and campaigning are mutually reinforcing. There is not necessarily an automatic correlation between media attention and the desired action being taken - ultimately some GM crops were approved despite there being no domestic market for them. During this period the sales of organic produce had reached new highs, marginalizing GM crops.

The brand

Brands are so ubiquitous that we are often not conscious of them, yet they are in many ways the symbolic framework of much of our lives. The symbols denoting and in part comprising the brand are intimately familiar to us, emblazoned on objects all around us. Yet beyond the brand symbol is a product, or set of products, that has certain functions and also certain personal or social meanings associated with it. The brand expresses particular values that we may or may not want to be associated with. This may feel like a necessity in that everything appears to be branded or we embrace particular brands as statements about ourselves and the world. Brands can become the commonplace markers of particular attitudes and dispositions of which in our social lives we are keen observers. Some of us like to imagine that we are not subject to the lure of the brand, viewing it as a controlled and controlling social phenomena which we would rather avoid. Yet, the reality of contemporary life is that it is more a matter of selecting between the brands available to us rather than rejecting them all (Klein, 2001).

Those who create and manage brands hold no illusions that they control them. They recognized long ago that brands have a social life that they cannot control, only guide. People when using a brand, sharing its meanings with others, become emotionally attached to it, and invest their hopes in it in a way that Arvidsson (2005) calls an 'ethical surplus'. This surplus arises through how people use the brand in their lives with other people; it is beyond the control or instigation of the brand creators and managers. We as consumers use brands to communicate to other people about our selves, forming 'brand communities' of different strengths and forms. For example, those who ride Harley Davidson motorbikes have clubs and meetings based on their shared love of the Harley; another group would be those who read and write letters to *The Times* or those who collect and restore vintage Ford tractors. This shared understanding starts with the brand but does not necessarily end there and can lead to further community developing. Brands are not, in this view, controlled by the corporations and indeed any company that tries to control them too closely is likely to see a revolt by the consumers.

The novelty of the organic 'brand' is that it is not controlled by a corporation but by a group of organizations representing the organic movement. As discussed above many of the values around the 'brand' are codified and open for public scrutiny. Yet, many of the other values of the brand are not set down in the same way but are rapidly apparent. The best example of this is the way in which the whole organic movement mobilized to resist the introduction of GM crops and foods into the UK. Genetically modified products were banned from organic production but the movement also moved to oppose their introduction into the UK food chain. At one point the campaign against GM was actively threatening the UK supermarkets with a consumer boycott and they rushed to remove GM products from their supply chains.

Compared to most corporate brands the organic brand is a volatile and difficult one to manage. For those seeking to profit from organic production, particularly larger businesses, the instability and values of the brand make it a difficult proposition. There are certainly profits to be made from organic produce, and the cachet of having an organic brand. Yet, the most obvious brand manager - the Soil Association - actively encourages its consumers to buy from box schemes, farmers markets and local shops. They are not the brand partners that the multiples are used to working with. Equally, there is no guarantee that those who certify organic production can fully control what it means as farmers, growers and consumers are constantly moving their aspirations. Hence, the growing call from consumers for ethically and fair traded goods that are also organic, pushing the standards to their limits in many instances.

Joining the dots

It might seem as if some of these concerns are disjointed but they do connect into one another in that the standards for organic farming are set by, and are broadly under the control of, a social movement which in turn holds broad sway over the brand under which those products are sold. Both the brand and standards are sites of contests between different actors; they are interconnected social arenas in which the meaning of organic is contested. It requires that the social movement is able to coordinate its arguments and activities between the two different venues. Yet in each slightly different tactics are needed, different forms of evidence and argument hold sway.

The different sites are places where the advance of organic land management can come quickly, and perhaps permanently, undone; the example of the United States perhaps illustrates this best. In the US the Federal Organic Standards have become an area where major corporations have attempted to undermine the international conception of organic agriculture, proposing for example that GM products could be organic (Lilliston and Cummins, 1998). This suggests that the movement in the US is not as able, as it is in Europe, to persuade either the political parties or networks of government. So the standards are not secure as a vehicle for building consumer trust, and many producers actively seek not to be involved with them. At the same time many in the organic movement are arguing that the other standards are so weakened that the organic as a brand is being captured by the retailers (Guthman, 2004); that organic as a brand has become hollowed out as the real concerns of agribusiness chase out the ideals of the organic movement.

There is little evidence of this occurring in the UK. The scientific evidence about the performance of organic farming in achieving the goals of improving the wider environment appears to have secured continued state support. As well as this the role that the organic movement played in the GM controversy and the strong support the public showed for it in the 'GM Nation' process would appear to have impressed on all of the major political parties that they need to support organic farming (AEBC, 2003). The multiple retailers have certainly benefited from the state support of organic farming in that they have been able to put some downward pressure on prices (Smith and Marsden, 2004). Yet, their rapid abandonment of GM products signals that they are probably not interested in moving into direct confrontation with the organic movement. Rather it would seem that they are content to attempt to ease their profile into that of the organic brand. The packaging of organic products and the point of sale materials now conjoin the term 'organic' with the name of the supermarket. Rather than the emblem of a challenge to the dominant views of agriculture and nature, the organic brand increasingly risks being seen as one of many contenders for space on the shelf or in the trolley.

Instead of breaking the rules of the game about the reckless commercial exploitation of nature, the organic brand may be drawn into becoming a competitor within the rules of the game that commodifies nature. This concern appears in recent controversy within the movement over the interpretation of the standards in a number of areas which has at times echoed external critics and at others opened new debates. Debates about aquaculture and the scale of poultry production under organic standards have garnered headlines, but there are also questions being raised within and outside of the organic movement about manure management, soil erosion and the sustainability of some inputs into organic systems. At the moment this appears to be leading to splinter organizations within the producer sector which may in turn be paralleled by the introduction of new certification schemes.

Through the standards, the 'brand' and the movement a much wider range of actors have become indirectly involved in sustainable land management. Whilst the motivations for selecting products are generally prompted by a variety of concerns - a confused web of regard for others and regard for the self - lurking within is the thought that the product selected is more sustainable. The new challenge for those looking to manage their land and produce food in an organic manner is to negotiate this tension between the aspirations of the end consumer, the strictures of the standards and the ethos of the social movement.

References

AEBC (2003) *GM Nation*. AEBC, London, UK.

Arden-Clarke, C. and Hodges, R. (1987) The environmental effects of conventional and organic/biological farming systems I. Soil erosion, with special reference to Britain. *Biological Agriculture and Horticulture* 4(2) 309-357.

Arden-Clarke, C. and Hodges, R. (1988) The environmental effects of conventional and organic/biological farming systems II. Soil ecology, soil fertility and nutrient cycles. *Biological Agriculture and Horticulture* 5(1) 223-287.

Arvidsson, A. (2005) Brands - a critical perspective. *Journal of Consumer Culture* 5(2) 235-258.

Blake, F. (1994) *The Organic Farming Handbook*. Crowood Press, Ipswich, UK.

Conford, P. (2001) *The Origins of the Organic Movement*. Floris Books, Edinburgh, UK.

della Porta, D. and Diani, M (1999) *Social Movements: An Introduction*. Blackwells, London, UK.

Dudley, N. (1991) *The Soil Association Handbook*. Optima, London, UK.

Guthman, J. (2004) *Agrarian Dreams. The Paradox of Organic Farming in California*. University of California Press, Berkeley, USA.

Howard, A. (1940) *An Agricultural Testament*. Oxford University Press, Oxford, UK.

Klein, N. (2001) *No Logo*. Penguin, London, UK.

Lampkin, N. (2005) Organic farming. In: Soffe, R. (ed.) *The Countryside Notebook*. Blackwell, London, UK, 181-198.

Lilliston, R. and Cummins, R. (1998) Organic vs 'organic': the corruption of a label. *The Ecologist* 28(4) 195-200.

Lobley, M., Reed, M., Butler, A., Courtney, P. and Warren, M. (2005) *The Impact of Organic Farming on the Rural Economy in England*. Final Report to Defra, CRR Research Report No.11. University of Exeter, Exeter, UK.

Miller, D. (2001) The poverty of morality. *Journal of Consumer Culture* 1(2) 225-243.

Moore-Colyer, R. (2004) Towards 'Mother Earth': Jorian Jenks, Organicism, the Right and the British Union of Fascists. *The Journal of Contemporary History* 39(3) 353-371.

Reed, M. (2002) Rebels from the Crown down: the organic movement's revolt against agricultural biotechnology. *Science as Culture* 11(4) 481-504.

Reed, M. (2004) *Rebels for the Soil: The Lonely Furrow of the Soil Association 1943-2000*. University of the West of England, Bristol, UK.

Reed, M. (2006) Turf wars: the UK organic movement's veto of GM. In: Holt, G. and Reed, M. (eds) *Sociological Perspectives of Organic Agriculture: From Pioneer to Policy*. CAB International, Wallingford, UK, 37-55.

Smith, E. and Marsden, T. (2004) Exploring the 'limits to growth' in UK organics: beyond the statistical image. *Journal of Rural Studies* 20, 345-357.

Soil Association (2003) *Food and Farming Report 2003 - Executive Summary*. Soil Association, Bristol, UK.

Soil Association (2005) Organic food sales growing by £2.3 million a week as local markets boom. *Press Release*. Bristol, UK.

Valentine, G. (1999) Eating in: home, consumption and identity. *The Sociological Review* 47(3) 492-524.

Part II

Communities of Information and Knowledge

Chapter 7

Just knowledge? Governing research on food and farming[1]

T. MacMillan

Introduction

Science and innovation are vital to food and farming. New technologies affect the way food is produced and how land is managed, sometimes for better and sometimes for worse. Research is needed to test new technologies and to monitor their effects on the environment, on health, on animal welfare and more besides. The relationship between science, policy and citizens has been severely shaken by a succession of controversies about risk regulation, new technology and public health. Food and farming research has been at the epicentre of this upheaval, in the wrangling over bovine spongiform encephalopathy (BSE) (van Zwanenberg and Millstone, 2003), Foot and Mouth Disease, genetically modified (GM) crops and, of late, obesity.

In 2000, a UK House of Lords committee reported that there was a crisis of public confidence in the governance of science and technology (House of Lords Select Committee on Science and Technology, 2000). A consensus amongst science organizations and government has crystallized around the committee's view that citizens have an undiminished enthusiasm for science, but are deeply sceptical that it is governed in the public interest. The policy response has been twofold: first, the evidence base for government decisions has been shored up with revised guidelines and processes for scientific advice (HM Government, 2000); and, second, 'science and society' programmes have bloomed, intended to promote public engagement in science.

A notable example of such public engagement aspirations was the *GM Nation?* debate of 2003, which attempted to engage non-scientists in deliberating science and technology on an unprecedented scale (GM Nation, 2003). Its success, however, has been widely questioned: not only has its methodology been criticized, but it also received from government neither the resources nor the thorough response that it warranted (Mayer, 2003; Radford, 2003; Horlick-Jones *et al.*, 2004). Nonetheless, it has helped to make food and farming research a central reference point for discussions of science and society across the board. The debate over GM crops was deeply divided and neither enthusiasts nor critics wish to see the same polarization repeated over other new technologies such as nanotechnology. Part of the problem was that public engagement was left too late, facing citizens and consumers with a stark choice: did they want this technology or did they not? A wide range of opinion is coalescing around the idea that citizens should engage in science further 'upstream', when choices remain open and research priorities are being set.

Public engagement should take place during research and development (R&D) rather than near the regulatory end of pipe. Organizations such as the Royal Society (2004; Royal Society and Royal Academy of Engineering, 2004) and Greenpeace (2004), at loggerheads over GM crops, agree on this. Demos and the Institute for Public Policy Research, two well-known think tanks, have published calls for upstream engagement (MacMillan, 2004; Wilsdon and Willis, 2004). An editorial in the science journal *Nature* argues that scientists should welcome public involvement in setting priorities for research, and it emphasizes that decision makers 'must explain why they choose to accept some pieces of advice and reject others' (*Nature*, 2004: 883). The UK government has pledged to promote public engagement upstream, in its 10-year *Science and Innovation Investment Framework*, and it has launched a funding programme called *ScienceWise* that is intended to make good on that commitment (HM Treasury *et al.*, 2004).

However, the post-*GM Nation?* consensus on upstream public engagement conceals profound differences in approach. For many government agencies and science organizations, public engagement is primarily about improving dialogue between scientists and non-scientists, sensitizing scientists to the concerns of non-scientists and vice-versa. The likes of Demos and Greenpeace, and the social scientists who introduced the idea of public engagement to policymakers, foresee a more radically reformed relationship between science and citizens. For them, the function of upstream engagement is 'to expose to public scrutiny the assumptions, values and visions that drive science' (Wilsdon and Willis, 2004: front cover). They would put questions about the purposes and framing of science and technology centre-stage in research decision making.

The gap between these two approaches is partly down to the place each ascribes to ethics in science. The former Prime Minister, Tony Blair, asserted that:

> Science is just knowledge. Science doesn't replace moral judgement ... It allows us to do more but it doesn't tell us whether doing more is right or wrong ... [W]ith scientific advance, we need greater moral fibre; better judgement; and stronger analysis of how to *use* knowledge for good not ill (Blair, 2002, emphasis added).

If science is 'just knowledge', a neutral reflection of reality, then the proper moment for public input is when that knowledge is applied to social ends. From this logic it would follow that 'upstream' public engagement in science, prior to its application, should be about communicating science rather than changing its direction.

By contrast, proponents of more thoroughgoing upstream engagement insist that science is not 'just knowledge' in the neutral sense that the former Prime Minister suggests. They argue that values and assumptions are built into the process of science and its products through the choice of research questions, through funding institutions and so on. This view of science as a thoroughly social practice, highlighted by Kuhn's (1962) work on scientific revolutions, has been elaborated in studies of how research agendas are formed, how scientific facts are produced and how science works (for a review see Jasanoff and Wynne, 1998). As well as describing this ethical content some writers make a normative claim: it is neither fair nor in the public interest for a small minority of citizens, mainly scientists, politicians and business people, to shape values and make assumptions that affect all of us through science. According to this view, enabling the public to play a constructive part in science and science policy is a precondition of a just research system and a just society (Wilsdon and Willis, 2004).

In December 2004, the Food Ethics Council (FEC) published a report, called *Just Knowledge?*, intended to contribute to these ongoing debates about the future of science, focusing on food and farming. The report explored what science would look like if it was not only technically rigorous, but also ethically and socially robust. This chapter briefly discusses some of the points that the report addresses in greater detail.

Ethics in science

'Ethics' has two common meanings. It can refer to the standards and values that define what is 'good' or 'right', or it can describe the study of those norms. In this chapter when I talk about ethics in food and farming science I mean the standards, values and framing assumptions that are built into the research system for food and farming. However some people, including the former Prime Minister and other senior policymakers, have said, in so many words, that they see ethical issues as entirely separate from science. They argue that science is neutral until it is 'used' or 'applied'. So my aim to explore the ethics of science is contentious or, at least, there is some confusion about the relationship between science and ethics. It is therefore worth spelling out, briefly, the three main premises on which this discussion of the ethics of science is based.

First, I see ethics as distinct from science. Science seeks to answer questions about what is, whereas ethics is more concerned with what ought to be. Science asks 'What is all this, and how does it work?', whilst ethics asks 'What should we do?' (Mepham, 2005: 5). You cannot determine what ought to be simply from knowing what the world is like.

However, second, ethics and science are far from separate. Indeed, each would be nonsense without the other. While ethical decisions do not simply follow from scientific knowledge about the world, we cannot decide what we ought to do without knowing about the situation we are in and how it might be changed. And the reverse is also true: science is underwritten by shared standards and values such as honesty and objectivity. It is therefore more helpful to treat ethics as a dimension of research and innovation, and as a set of questions to ask about those activities, than as a separate set of concerns.

My third premise is that the ethical dimension of research is not always plain to see. Indeed, sometimes the ethical content of science is emphatically denied. In particular, the rigour of the scientific method is sometimes mistakenly assumed to expel ethical values and social assumptions from scientific knowledge. But just because scientific knowledge is relatively reliable – in contrast to an uncorroborated news report, for instance, we might be able to repeat a scientific experiment and get the same result – the ethical and social factors that shape research questions, that steer science funding or that inform accepted benchmarks for statistical significance do not diminish in importance. Yet while few practising scientists consistently claim that science is separate from ethics or free from value assumptions, social scientists have described how scientists, politicians and business people have used more flexible claims about these issues to manage their responsibilities for science in society. Studies have outlined how through strategic 'boundary work' (Gieryn, 1999) a line is drawn between science and ethics when it suits, for example in defending the expertise of scientists compared with 'lay people', but erased at other times, as in discussion of funding pressures on scientists.

Once we recognize that science has an ethical dimension, it no longer makes sense to think of robust science as being free from values. Insisting that ethics is separate from science - saying that science is 'just knowledge' - simply shields the framing assumptions of scientists and decision makers from scrutiny by other scientists and citizens. It does not eliminate that ethical content.

This chapter argues that science contains assumptions about society and social values. My main criterion for an ethically robust research system is that these assumptions should be open to social scrutiny. Just as the famous philosopher of science Karl Popper (1979) exhorted scientists to expose their technical and empirical claims to challenges by others, so should the value assumptions science makes on behalf of society be open to question. Such exposure, I argue, will create a more robust research system.

Farming and food research - an ethical overview

Statements about the relationship between science and ethics in general may seem abstract, but they are crucially important to what scientists, policymakers and farmers actually do. For a start, these abstract questions about ethics and values frame policies on farming and food research. For example, open meetings of the Science Advisory Council, Defra's senior advisory body on science established in February 2004, often include considerable discussion about values in science and about public involvement in decision making (Defra, 2004). More profoundly, perhaps, there is a sense in which research on food and farming is a moral failure. It is a very long way from satisfying four commonly held ethical principles, which can be expressed in the form of four questions:

1. Is it maximizing the benefits of science and technology to society and the environment?

For the most part it is not. Science and technology are not orientated towards addressing the greatest social and ethical challenges in food and farming. In particular, the research system puts profit and productivity, in a narrow sense, before hunger alleviation. Yet around 840 million people in the world are severely malnourished, and many more survive on diets that may be technically adequate but are very poor by UK standards. Although new technology is not necessary to eliminate hunger or malnutrition, according to the United Nations (UN) Food and Agriculture Organization (FAO), much research that could help goes un- or under-funded (FEC, 1999a; FAO, 2003; FEC, 2003).

2. Is it minimizing the harmful side-effects of science and technology?

Again, no. Heavy investment is directed towards science and technology that demonstrably harms people, animals or the environment. For instance, recombinant bovine somatotrophin (rbST), a synthetic hormone used in the USA to boost milk production, is banned in the European Union (EU) because it was found to harm cows (FEC, 1999b).

3. Is it enhancing people's autonomy and freedom of choice?

Food processing technologies have certainly enabled manufacturers to create an expanding array of differentiated products but it has been from a narrow range of ingredients, notably fat, sugar and salt. So the choice of products available to consumers has increased but, as the availability of and information about other foods has declined by comparison, consumers' dietary self-determination has diminished. This trend has been implicated in the global 'epidemic' of diet-related disease (FEC, 2001; POST, 2003).

4. Finally, is the research system just and fair?

Not at all. Key rule-making processes that shape farming and food research are deeply unfair. For example, low-income countries and marginal rural communities were poorly represented during negotiations to create international rules on intellectual property, which affect how different kinds of knowledge are valued and who benefits from their exploitation (CIPR, 2002; FEC, 2002).

Policy context and priorities

As described in the introduction, the context for this discussion about the ethics of science is a crisis of public confidence in the ways that science and technology are governed, the effects of which are reverberating through policy. The first policy response has been to strengthen the scientific advice given to decision makers, with increased emphasis on evidence-based policy and new guidelines for government advisers. The second response has been to promote 'upstream' public engagement in science.

I agree that greater and earlier public involvement in research and research policy is needed. However, I also think the effect of initiatives to promote upstream public engagement will be cosmetic unless they are backed up with *other* changes to the ways in which science and technology are governed.

First of all, policies on science and technology must be made more consistent. The policies that affect research on food and farming are not very well joined-up at the moment. In particular, the government's efforts to promote innovation and business R&D have been insulated from its commitment to upstream public engagement (MacMillan, 2004). So we have a situation where, on the one hand, the government treats public engagement as a *brake* on scientific progress, albeit a necessary one. By contrast, on the other hand, non-scientists representing business are routinely involved in decisions about science, on the assumption *they* will help drive science and technology *forward* (HM Treasury *et al.*, 2004). Current discussions about public engagement take place in the context of a real mismatch between the ways that different groups of stakeholders are treated by policymakers. Different *methods* of engagement may of course be appropriate but different *rules* of engagement are certainly not.

Second, the focus on wealth creation, in a narrow sense, as the primary objective of science policy conflicts with the government's commitment to sustainable development (HM Government, 2005). Sustainable development encompasses economic aims but places equal emphasis on social and environmental goals - it starts from the recognition that some things which are commercially profitable are bad for people and the environment. Thus there is a need for the principles of sustainable development to inform policy on science and technology.

The research system must also be made more accountable. Public engagement should be treated as a complement to political representation in decision making, not as a substitute for it. Policy advice should be transparent and independent, and it should open up the possibilities available to decision makers. So, if a public body gets advice from experts and from the public on a topic such as obesity research or animal genetics, it is not obliged to follow either but it should explain its decision in response to both.

Finally, science and technology should be governed more fairly - in other words, with a greater concern for social justice. The privatization of public sector research is concentrating decision making power and research resources in the hands of corporate stakeholders, at the expense of other citizens, of science workers and of farming communities. A serious restructuring of decision making and a radical redistribution of research resources are both preconditions of a just research system. The UK government cannot achieve this alone but the Food Ethics Council report (2004) finds much it *could* be doing that it isn't.

Towards a just research system

The Food Ethics Council does not pretend to have all the answers but its 2004 report suggests priorities for change in four areas. In research and education, it proposes that greater efforts should be made to encourage scientists to deliberate on the social consequences of their work. Funding bodies such as the Biotechnology and Biological Sciences Research Council can help by supporting grant applicants that take this aspect of their research seriously, and by providing training and support.

In risk regulation, social and ethical concerns have been systematically excluded from consideration. In some cases it is perceived that addressing them would contravene international trade agreements. The UK government should press to change those rules. The report also recommends it establishes a clear responsibility for participatory technology assessment, drawing on the experience of other European countries.

Policies on science need to be revisited, not only looking inward at the way policymakers use science, but also outward at the effects of science policy on other areas of government, on society and on the environment. At the moment, government interventions in research and innovation focus on boosting UK economic competitiveness. Adopting sustainable development as a primary objective for science policy instead, as the Food Ethics Council's report recommends, would ease conflicting pressures on public research. It would also imply a different approach to innovation, predicated on the precautionary principle, perhaps aiming to increase the range of technological options rather than trying to 'pick the winners'.

Finally, the Food Ethics Council (2004) recommends that the government helps to redistribute control of the resources for doing research. This is partly a case of reinvesting in public sector research, safeguarding the independence of regulatory agencies, and improving capacity to monitor *how* public and private research organizations are spending their money. But it is also about ensuring incentives for knowledge creation are in the public interest. Policies governing the exploitation of intellectual resources should aim to reward collective creativity, to combat the use of patents to block R&D, to alleviate commercial pressures on public research, and to strengthen the kinds of informal knowledge that develop outside of the business and scientific communities.

I believe it is possible to meet these challenges. This government has demonstrated a clear commitment to science and *also* ratcheted up its public engagement programmes. The time has come to combine the two, and for the government to take a lead in promoting science and technology that are not only technically rigorous and of economic value, but are also socially, environmentally and ethically robust.

Notes

1. This chapter is based on a report of the same name published by the Food Ethics Council (FEC) (2004) and some of the text is excerpted from the report. The full report is available from http://www.foodethicscouncil.org

References

Blair, T. (2002) *Science Matters*. Speech 23/5/2002. London, UK. [Accessed 2007] Available from http://www.number-10.gov.uk/output/Page1715.asp

Commission on Intellectual Property Rights (2002) *Integrating Intellectual Property Rights and Development Policy*. CIPR, London, UK.

Defra (2004) New Science Advisory Council for Defra. *Defra News Release*, Ref. 34/04. 3 Feb. [Accessed Jan 2007] Available from http://www.gnn.gov.uk/content/detail.asp?ReleaseID=107 174 &NewsAreaID=2&NavigatedFromSearch=True

FAO (2003) *The State of Food Insecurity in the World 2003*. Food and Agriculture Organization, Rome, Italy.

FEC (1999a) *Novel Foods: Beyond Nuffield*. Food Ethics Council, Southwell, Nottinghamshire, UK.

FEC (1999b) *Drug Use in Farm Animals*. Food Ethics Council, Southwell, Nottinghamshire, UK.

FEC (2001) *After FMD: Aiming for a Values-Driven Agriculture*. Food Ethics Council, Southwell, Nottinghamshire, UK.

FEC (2002) *TRIPs with Everything? Intellectual Property Rights and the Farming World*. Food Ethics Council, Southwell, Nottinghamshire, UK.

FEC (2003) *Engineering Nutrition: GM Crops for Global Justice?* Food Ethics Council, Brighton, UK.

FEC (2004) *Just Knowledge: Governing Research on Food and Farming*. Food Ethics Council, Brighton, UK.

GM Nation (2003) *GM Nation? Findings of the Public Debate*. DTI, London, UK.

Gieryn, T. (1999) *Cultural Boundaries of Science: Credibility on the Line*. University of Chicago Press, Chicago, USA.

Greenpeace (2004) Response to HM Treasury (HMT), Department for Education and Skills (DfES) and Department of Trade and Industry (2004) *Science and Innovation Investment Framework 2004-2014*. HMSO, London, UK. [Accessed 2004] Available from scienceframework@hm-treasury.gov.uk

HM Government (2000) *Guidelines on Scientific Analysis in Policy Making*. Second edition. Office of Science and Technology, London, UK.

HM Government (2005) *Securing the Future: Delivering UK Sustainable Development Strategy*. Cm. 6467. HMSO, London, UK.

HM Treasury (HMT), Department for Education and Skills (DfES) and Department of Trade and Industry (2004) *Science and Innovation Investment Framework 2004-2014*. HMSO, London, UK.

Horlick-Jones, T., Walls, J., Rowe, G., Pidgeon, N., Poortinga, W. and O'Riordan, T. (2004) *A Deliberative Future? An Independent Evaluation of the GM Nation? Public Debate about the Possible Commercialisation of Transgenic Crops in Britain, 2003*. Understanding Risk Working Paper 04-02. UEA, Norwich, UK.

House of Lords Select Committee on Science and Technology (2000) *Science and Society*. Third report, 23 February. HMSO, London, UK.

Jasanoff, S. and Wynne, B. (1998) Science and decision-making. In: Rayner, S. and Malone, E.L. (eds) *Human Choice and Climate Change: the Societal Framework*. Vol.1. Battelle Press, Columbus, Ohio, USA, 1-87.

Kuhn, T. (1962) *The Structure of Scientific Revolutions*. University of Chicago Press, Chicago, USA.

MacMillan, T.C. (2004) *Engaging in Innovation: Towards an Integrated Science Policy*. IPPR, London, UK.

Mayer, S. (2003) *GM Nation? Engaging People in a Real Debate?* GeneWatch UK, Buxton, UK.

Mepham, B. (2005) *Bioethics: an Introduction for the Biosciences*. Oxford University Press, Oxford, UK.

Nature (2004) Going public. *Nature* 431, 883.

POST (2003) *Improving Children's Diet*. Parliamentary Office of Science and Technology, London, UK.

Popper, K.R. (1979) *Objective Knowledge*. Revised edition. Oxford University Press, Oxford, UK.

Radford, T. (2003) Scientists complain GM debate was mishandled. *Guardian*, 1 November. [Accessed 2004] Available from http://education.guardian.co.uk/higher/news/story/0,,1076645,00.html

Royal Society (2004) Response to HM Treasury (HMT), Department for Education and Skills (DfES) and Department of Trade and Industry (2004) *Science and Innovation Investment Framework 2004-2014*. HMSO, London, UK. [Accessed 2004] Available from scienceframework@hm-treasury.gov.uk

Royal Society and Royal Academy of Engineering (2004) *Nanoscience and Nanotechnologies: Opportunities and Uncertainties*. Royal Society and Royal Academy of Engineering, London, UK.

van Zwanenberg, P. and Millstone, E. (2003) BSE: a paradigm of policy failure. *Political Quarterly* 74(1) 27-37.

Wilsdon, J. and Willis, R. (2004) *See-through Science: Why Public Engagement Needs to Move Upstream*. Demos, London, UK.

Chapter 8

Agricultural advisers and the transition to sustainable soil management in England: a focus on agronomists' understanding of soil

J. Ingram

Increasing demands on advisers

The agricultural adviser has always been an important and influential actor on the farm both in terms of delivering specialist technical advice in support of food production (Dancey, 1993) and more recently in assisting in the implementation of agri-environment policy (Cox *et al.*, 1990; Burton *et al.*, 1999; ECOTEC, 2000; Juntti and Potter, 2002). Indeed of all methods of communication, the individual farm visit by an adviser has remained one of the most powerful and effective and most valued by the farmer (Jones *et al.*, 1987; Eldon, 1988; Fearne, 1991; Angell *et al.*, 1997). Today their role is more important than ever in supporting farmers as they struggle to meet the challenges and uncertainties of changing policy, regulations, technology and markets. Although mechanisms that promote farmer group learning are becoming increasingly popular (Roling and Wagemaker, 2000), farmers still value interpretation of information at the farm level, they are also increasingly reluctant to share knowledge with their peers as each tries to retain a competitive advantage (Angell *et al.*, 1997). Advisers therefore remain an essential component of the Agricultural Knowledge System (AKS) in England, figures for those providing both commercial and public-good messages are testament to this (Dauven *et al.*, 1998; Dauven and Crabb, 1999), indeed 96% of farmers in the UK used an adviser in 2002 (JT Research Agribus, 2003). However, whilst it is accepted that advisers are an important and influential actor on the farm, little is known about the role different types of advisers can play in the transition to more sustainable agriculture.

Within England, new policy and research priorities in agriculture which emphasize sustainable production and environmental protection rather than productivity, together with farmers' changing requirements for advice in a more competitive and restructured industry, have inevitably impacted on the role of agricultural advisers and brought new challenges. This is particularly the case in the context of soil management where recent developments in England herald a new era of policy concern (Royal Commission on Environmental Pollution, 1996; Defra, 2004a; Environment Agency, 2004a; 2004b). More recently the requirements for farmers to prepare a Soil Protection Review (SPR) as a condition of receiving the Single Payment (Defra, 2006), to complete a soil risk assessment as part of their Farm Environment Record to qualify for Environmental Stewardship Scheme payments and to prepare a Soil Management Plan (SMP) as an option for the Entry and Higher Level scheme all place new demands on advisers. The introduction of catchment sensitive farming associated with the implementation of the EC's Water Framework Directive represents a further challenge (Defra, 2004b).

To achieve implementation of these policies, farmers are being urged to undertake a suite of 'best management practices' for soil (see MAFF, 1998; ADAS *et al.*, 2000a; 2000b; Environment Agency, 2001; Defra, 2005; UK Soil Management Initiative, 2005). These practices are based on a number of fundamental principles including: maintenance of soil structure through enhanced soil organic matter content and careful cultivation to avoid compaction, overworking and runoff; as well as the management of soil as a buffer for nutrients by targeting artificial and organic fertilizers effectively. Farmers can lack familiarity with, and experience of, such complex practices which can be non-prescriptive and demand attention to detail (OECD, 2001). As such there is a clear requirement for more information and on-farm advice to support them in using these practices, particularly given the heterogeneity of farm soils and the individuality and locally specific nature of farm based planning that is needed. This has also been demonstrated for other knowledge demanding practices which provide environmental and soil benefits such as integrated farming systems (Park *et al.*, 1997; Morris and Winter, 1999), reduced tillage (Tebrugge and Bohrnsen, 2001; Coughenour, 2003), organic farming (Burton *et al.*, 1999) and sustainable soil management (Ingram, 2008). In recognition of this, policy development is based on an understanding that advice to farmers is crucial to achieving sustainable farm practices (Garforth *et al.*, 2003; Defra, 2004a; Defra, 2004c). This chapter aims to explore how advisers, agronomists in particular, are equipped to provide such advice.

Evidence has shown that some advisers have been very effective in assisting policy implementation by providing advice on conservation (Cox *et al.*, 1990); agri-environment schemes (Cooper, 1999), organic farming (Burton *et al.*, 1999) and pollution prevention measures (ECOTEC, 2000). Many argue, however, that it is the commercial advisers, the agronomists or crop consultants, who establish regular one-to-one contact with farmers and are most valued and trusted, who are best placed to relay environmental protection messages to farmers (Jones *et al.*, 1987; Winter, 1995; Gasson and Hill, 1996; Archer, 2001; Dampney *et al.*, 2001). Whether they are fit for purpose in providing such advice is, however, uncertain. It has been claimed that commercial advisers are ill-equipped to promote public good messages; that they promote intensification (Eldon, 1988) and that consultancy is not a substitute for extension (Gasson and Hill, 1996). Others consider that they perceive and treat environmental practices as a constraint, rather than an objective of farming practices and that they still hold what Wilson (2001) calls 'productivist modes of thinking'. This is related to what some have described as a 'structural inertia' in providers due to market orientation of advice services and the legacy of production-oriented advice which is enduring within all advisory services (Winter, 1996; 1997; Curry, 1997). Proponents of this view also claim that commercial advisers' environmental skills are poorly developed (Winter, 1995; Curry, 1997). However, more recent research suggests that commercial advisers can play a positive role in facilitating best management practice. Although Marshall's (2002) survey of agronomists' environmental skills reflected different levels of knowledge and experience, it found that a significant number were already giving sound environmental and good agricultural practice advice. There is further evidence as well of the agronomy industry in general responding to new policy signals by improving their 'green credentials' (*Farmers Weekly,* 2006) and recognizing the need for effective soil management advice (*Farmers Weekly,* 2004a).

Methodology

The research on which this chapter is based combined an extensive postal questionnaire survey of a range of agricultural advisers and semi-structured interviews with a selected group of advisers. The proliferation of advisers in the AKS in England following privatization of the state advisory service, ADAS, in 1997 has resulted in a complex and diverse advisory community (Winter, 1995; Dampney *et al.*, 2001; Winter *et al.*, 2001; Garforth *et al.*, 2003). As soil best management practices are not specialist technologies but integral to all farm practices they become the potential remit of a wide range of arable, environmental protection and conservation advisers offering commercial and public good advice. Consequently the target population for the questionnaire survey was all agricultural advisers in England providing one-to-one advice which might have some impact on soil management decisions on the farm.

Synthesis of information from a number of recent reviews of advice provision (Archer, 2001; Dampney *et al.*, 2001; Winter *et al.*, 2001; Marshall, 2002) provided a sampling frame, and 304 questionnaires were sent to individual advisers in four categories as follows: conservation advisers in the Farming and Wildlife Advisory Group (FWAG) and the National Trust; agri-environment scheme advisers in Defra's Rural Development Service (RDS) (including advisers in the Organic Advisory Service); independent agronomists (crop consultants) identified through the Association of Independent Crop Consultants (AICC) and British Institute of Agricultural Consultants (BIAC), and ADAS advisers. A fifth category, commercial agronomists working for agro-chemical and seed distributor companies, were approached separately through the Fertilizer Advisers Certification and Training Scheme (FACTS). Due to the different approaches to identifying potential advisers, non-probability sampling was used to target certain sectors. However, in most cases all advisers providing one-to-one advice potentially affecting soil management were approached ensuring that as far as possible the full population was contacted. In total 163 questionnaires were returned with an average response rate of 40% for the first four categories. It was not possible to estimate a response rate for the FACTS respondents.[1]

Quantitative data from the questionnaires were complemented by qualitative data derived from semi-structured interviews with a range of advisers from the same categories. These advisers were loosely associated with two soil management initiatives, in England, the UK Soil Management Initiative (SMI) and the Landcare Partnership in the Upper Hampshire Avon catchment. These promote a range of best management practices including targeted use of nitrogen (N), use of nutrients within manure and good cultivation practice to improve soil structure. Sampling was undertaken using lists of advisers participating in the initiatives, local directories of advisers and the snowball technique. Sixty-four advisers were selected for interview based on geographical location (where the initiative was in a distinct area), involvement, and or, likely interaction with the initiative. These interviews provide different accounts and interpretations that both add to the multiple sources of evidence and assist in the triangulation of quantitative material.

Results

In the analysis, although results are presented for all adviser categories, there is an emphasis on agronomists.[2] This is justified on the basis that, of all advisers, agronomists have most opportunity to influence farmers' soil management behaviour through their regular on-farm contact and their provision of arable advice.

Advisers' understanding of best management practice for soil

Questionnaire data are used to analyse advisers' observations of soil degradation, their use of soil best management practice guides and the extent of soil best management practice recommendations made. Interview data are then used to explore in more depth agronomists' perspectives in the context of cultivation.

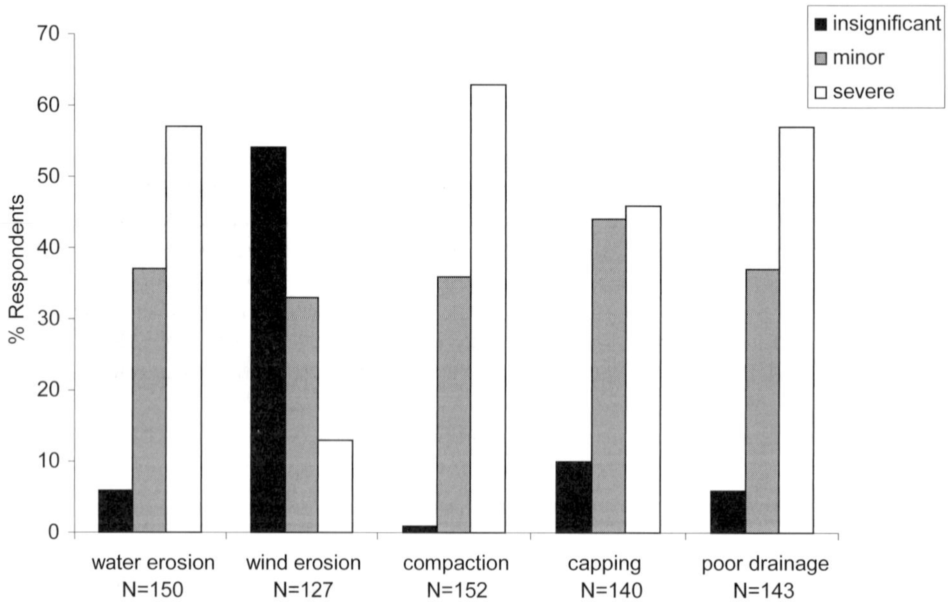

Fig.8.1: Advisers' response to the question: 'Indicate the severity of any incidences of degradation which you have observed in the course of your work over the last 2 years which can be attributed to inappropriate land use or poor soil management' (where degradation was defined in the questionnaire as any short or long-term damage to soil that impairs its functions)

Advisers' observations of soil degradation

More than 50% of the questionnaire respondents had observed what they described as 'severe' water erosion, compaction, capping and poor drainage attributed to inappropriate land use over the last 2 years (Fig. 8.1). Obviously some soils are more vulnerable to degradation than others and, as this data cannot be related to any particular soil type, care must be taken in its interpretation. Many respondents stressed the localized occurrence of degradation and the effect of extreme rainfall. Agronomists working in the Upper Hampshire Avon catchment distinguished the more erosive Greensand soils from the chalk soil but pointed out that rainfall and cultivation practice are important factors as well:

> I think everyone's very conscious about erosion ... it's very noticeable on roads. It's a big thing particularly on the Greensand. Some of the gullies in Hampshire are a foot wide and a foot deep, they all got ploughed in and with more rain it got washed out again, unbelievable, quite extraordinary (Distributor agronomist R).
>
> The chalk doesn't normally run but last year it did, there was classic over working, ploughing, then power harrowing and drilling it, then with showers or heavy rain the soil was washed away (Distributor agronomist RW).

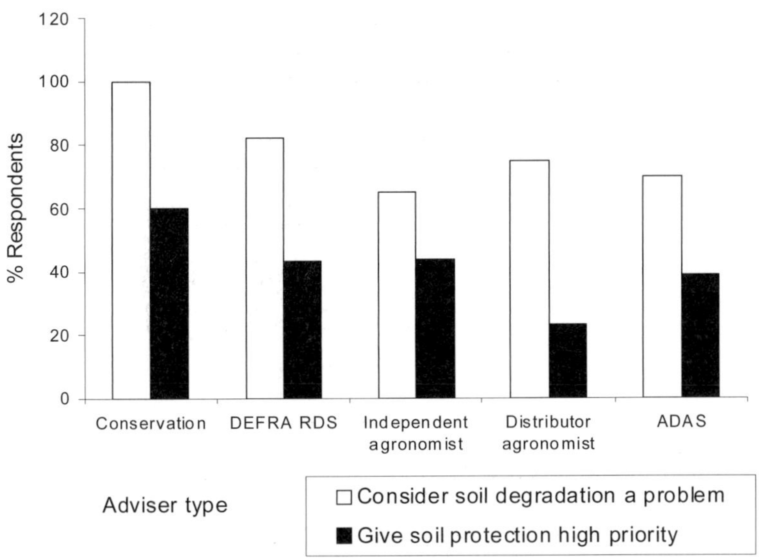

Fig.8.2: Adviser concern about soil degradation and priority given to soil protection (where soil protection refers to protection in its broadest sense where long-term damage to soil is avoided)

In accordance with these observations of soil degradation, more than 65% of all respondents thought that soil degradation was a problem in English agriculture (Fig. 8.2), again with qualifications about its localized and irregular nature. Approximately the same proportion of agronomists (independent 65%, distributor 75%) as ADAS advisers (70%) thought it a problem compared to a higher percentage of non-arable advisers, RDS (83%) and conservation advisers (100%). Advisers who considered that no problem exists, tended to dismiss concerns about soil degradation, one agronomist for example claimed that 'soil structure is better now than at any time in the last 20 years'. However, despite these different view points, the questionnaire and interview data do give an impression that many advisers are observing a range of soil conditions which they interpret as symptomatic of degradation.

Adviser use of soil best management practice guides

The extent to which advisers use guides available to support best management practice decisions was explored for the following: the Soil Code (MAFF, 1998), the Controlling Soil Erosion Manual (MAFF, 1999a; 1999b; Defra, 2005) and Soil Survey maps and bulletins. The Soil Code is a Defra publication sent to all farmers and owning and abiding by the Code is a requirement for most farm assurance and agri-environment schemes. The manual is designed to help land managers assess and control erosion risk. Figures from the questionnaire suggest that use of the Soil Code is reasonably high, with 11%, 55% and 34% respondents respectively using it never, sometimes and always; that is the majority of respondents use it at least sometimes. Figures (Table 8.1) for individual adviser types demonstrate that RDS, ADAS and conservation advisers use the Soil Code most frequently, whilst agronomists use it marginally less, although 31% of distributor agronomists never use it. Fewer advisers use the Controlling Soil Erosion Manual; the data show that 58% of all respondents never use it, with 37% sometimes using it and only 5% always using it.

Table 8.1: Adviser use or recommendation to farmers of the Soil Code

	% of respondents within each adviser category				
	Conservation	Defra RDS	Independent agronomist	Distributor agronomist	ADAS
Never	3	5	14	31	5
Sometimes	50	43	61	46	59
Always	47	52	25	23	36
No. of valid respondents	32	21	71	13	22

These figures need to be set against the 57% of all respondents who observed severe incidents of soil erosion (see Fig. 8.2). Compared to the Soil Code, the manual is a less established document, addresses more specific problems and soils and is not a formal requirement for schemes. ADAS and conservation advisers made greatest use of this manual (Table 8.2). The questionnaire data also show that nearly half of all respondents (conservation (56%), independent agronomist (44%), distributor (54%) and ADAS (61%)), with the exception of Defra RDS advisers (21%), refer to soil survey maps and accompanying bulletins which provide technical and management data.

Table 8.2: Adviser use or recommendation to farmers of the Controlling Soil Erosion Manual

	% of respondents within each adviser category				
	Conservation	Defra RDS	Independent agronomist	Distributor agronomist	ADAS
Never	25	64	75	62	45
Sometimes	72	29	22	23	50
Always	3	7	3	15	5
No. of valid respondents	29	14	67	13	22

Table 8.3: Extent of adviser training and FACTS registration

	% of respondents within each adviser category				
	Conservation	Defra RDS	Independent agronomist	Distributor agronomist	ADAS
Training	77	80	90	69	81
No. of valid respondents	31	20	73	13	22
FACTS registered	12	34	79	100	87
No. of valid respondents	32	23	73	13	23

The extent of advisers' training and FACTS registration

A large proportion of advisers from all sectors claimed to have undertaken training to assist them with soil best management practice recommendations, as Table 8.3 shows. Although soil management training is not given *per se*, it is integral to the FACTS courses and to the training provided by ADAS as part of Defra's campaigns to promote the efficient use of fertilizers and manures and the control of erosion. Advisers also attended workshops and presentations, for example, those run by the UK SMI, as well as in-house courses within commercial firms and consultancy groups.

Advisor recommendations of soil best management practice

More than 70% of advisers have recommended carefully timed cultivations, minimum tillage and buffer strips,[3] and 58% recommended low compaction machinery and early autumn sowing - fewer recommended cover crops and contour ploughing and the more specialist practices such as precision farming and bi-cropping (Table 8.4).

Table 8.4: Adviser recommendations of soil best management practices in the last 2 years

Cultivation recommendations	%	Nitrogen recommendations	%	Anti-erosion measures	%	Others	%
Carefully timed cultivation n=157	82	Targeted N n=157	77	Buffer strips n=155	73	Green manures n=157	42
Minimum tillage n=157	70	Manure N value n=157	82	Contour plough n=156	24	Bi-cropping n=157	10
Low compaction machinery n=157	58			Cover crops n=155	52	Precision farming n=157	28
Early autumn sowing n=155	58					Permanent grass n=155	48

Figure 8.3 shows that advisers from all sectors recommend these practices to a similar extent apart from RDS advisers, who in all cases, except buffer strips, recommend these practices the least. Significantly, as a group, agronomists recommend these practices to almost the same extent as ADAS advisers. Low figures for RDS advisers are most likely because these are practices integral to arable cropping but not to habitat or agri-environment scheme management, although surprisingly conservation advisers claim to recommend all practices to the same extent as arable advisers.

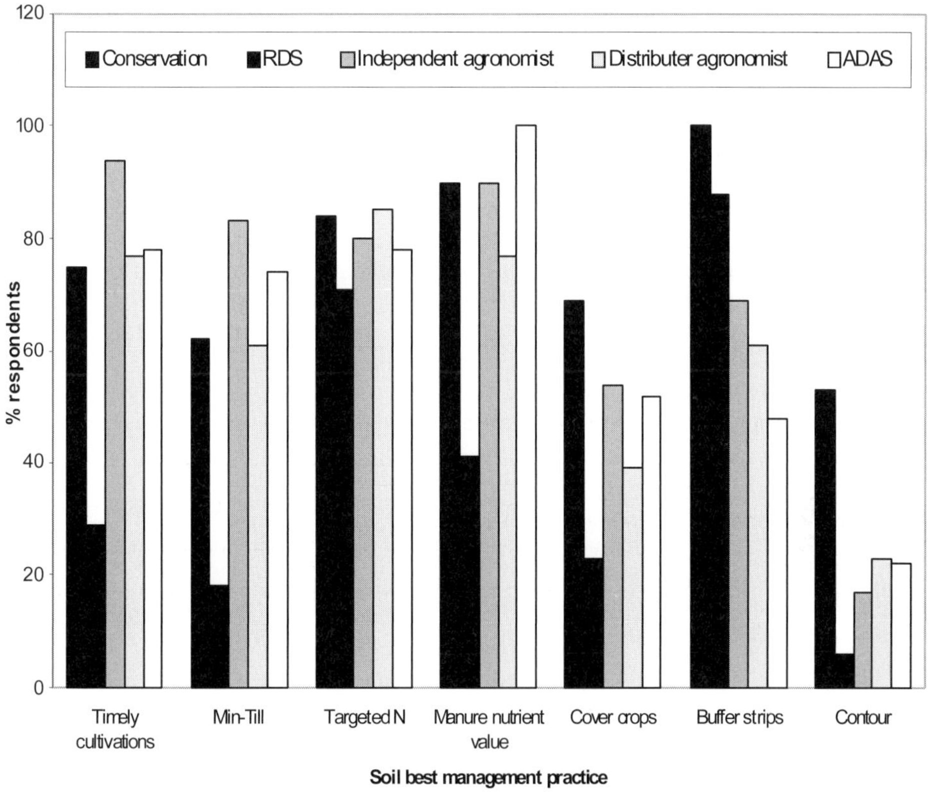

Fig.8.3: Adviser recommendations of soil best management practices in the last 2 years

Agronomists and soil management: a focus on cultivation

Most agronomists interviewed regard soil best management practice as sound practice, and believe they are already promoting it as part of their day-to-day advice. Many are increasingly recognizing the dual benefits to the environment and the farm business of best management practice and emphasized the importance of soil as the basic material for farming, as one distributor agronomist (TB) remarked: 'If you're an agronomist the first thing you should be doing is looking at the soil, picking it up and feeling it'. For many agronomists however soil is not singled out as a resource that needs particular attention but its management is seen as integral to the whole farming system. As such, soil is not accorded any higher priority than any other agronomic factor. Indeed Figure 8.2 demonstrates how, despite concerns about degradation, soil protection is not given comparable priority. Comments from those outside the agronomy community suggesting that soil management is outside the agronomist's remit, support this:

> They [agronomists] are very good but most take little interest in soils. They expect the farmer to look after the soil and cultivations and they look after the plant and hope it has got some roots (ADAS adviser).

Using the interviews to further explore the nature of agronomists' knowledge about soil revealed that, although they demonstrate use of guidelines and recommend practices such as appropriate cultivations and minimum tillage, their understanding of how to implement these practices is limited. Decisions about cultivations are seen as some of the most important on the farm but the interviews show that this is an area where agronomists are least involved and have minimum confidence and expertise. Many regard the absence of hands-on experience of cultivation as a big gap in agronomists' knowledge. The emphasis on needing to be a practitioner to understand soil capability in respect of cultivations is echoed in many interviews, as this comment suggests:

> A lot of people I know who work in the soils field and agronomy advisory field are actually very weak on soils, on hands-on soils. They really are. They just haven't got their minds around it. One of the fundamental things is timeliness and soil moisture. The only way you'll understand about soil capability and timeliness is by doing it and doing it for quite a while and that's the problem, these people, they don't actually do it (Adviser R).

Others agree with this view suggesting that advisers' support for farmers' cultivation decisions is weak:

> There has got to be a much better informed adviser group in this country able to look at farms and interpret and ensure they are using the right equipment and discuss with farmers in a knowledgeable way which way their cultivation should be going. At the moment an awful lot of farmers suck it and see (Consultant J).

Agronomists recognize these limitations, they acknowledge that farmers have the hands on practical skills and that they have to tread carefully in areas where their competence might be questioned, as this agronomist observes:

> I'm an advisory agronomist who has a lot of technical information at my fingertips but I'm not a practical farming manager and we are stepping over the borders here a little bit. Most farmers are experts in machinery or like to think they are. They will listen to me but what they say when I've gone, I don't know (Independent agronomist J).

This suggests that despite agronomists' recognition and concern about soil degradation, their use of guides and recommendations, their skills in the most critical aspect of soil management in the field, cultivation, is limited. This has significant implications for sustainable soil management.

Advisers' understanding of nutrient management

This section first examines the extent to which advisers use tools and recommend practices with respect to nutrient management using the questionnaire data, and then goes on to explore in more detail the nature of their understanding of nutrient management with a focus on agronomist recommendations.

Adviser use of tools that assist them with nutrient management

A number of tools are available to help growers and advisers predict N fertilizer requirements. These include guides like RB209 Fertilizer Recommendations for Arable and Horticultural Crops which is a well established document developed by ADAS (MAFF, 2000), Decision Support Systems (DSS) such as PC operated Fertilizer Recommendation Systems (FRS), soil analysis and tools that allow the nutrients supplied by manure to be estimated.

The extent to which advisers use these is shown in Tables 8.5, 8.6 and 8.7. The same pattern emerges for the use of each tool with arable advisers using all these tools to a much greater extent than the conservation and RDS community, as would be expected with tools which support arable decisions. A higher proportion of ADAS advisers use these tools than agronomists, which is not surprising given that many were developed by ADAS. Figures suggest however that agronomists are using these tools to a considerable extent as well.

Table 8.5: Adviser use of soil management tools when advising farmers

	% of respondents within each adviser category				
	Conservation	Defra RDS	Independent agronomist	Distributor agronomist	ADAS
Soil analysis	0	0	3	8	9
DSS	9	5	22	30	n/a
MANNER	12	23	53	46	74
No. of valid respondents	32	19	72	13	23

A very small proportion of advisers use soil analysis which may be due to the cost that has to be passed onto the farmer. Alternatively, advisers may lack the skills and understanding in taking soil samples and interpreting the analysis. Limited use of DSS, which assist on-farm decisions about nutrient management, also suggests that advisers have yet to embrace such tools. MANNER (MAnure Nitrogen Evaluation Routine), which is a PC tool developed by ADAS to assist farmers and advisers accommodate manure in their nutrient budgets, however, is popular for estimating the nutrient contributions made by manures. This together with use of the Managing Livestock Manure booklets (ADAS *et al.*,

2000a; 2000b) among arable advisers suggests that they are integrating nutrients from manure into arable fertilizer schedules. Both the questionnaire data and interviews revealed that the use of the fertilizer guide RB209 is widespread among all advisers, with 95% of all respondents from the arable sector using it at least sometimes.

Table 8.6: Adviser use of Managing Livestock Manures booklets when advising farmers

	% of respondents within each adviser category				
	Conservation	Defra RDS	Independent agronomist	Distributor agronomist	ADAS
Never	18	50	16	46	4
Sometimes	78	33	62	31	65
Always	4	17	22	23	31
No. of valid respondents	27	18	69	13	23

Table 8.7: Adviser use of RB209 when advising farmers

	% of respondents within each adviser category				
	Conservation	Defra RDS	Independent agronomist	Distributor agronomist	ADAS
Never	39	44	7	8	0
Sometimes	57	22	40	38	30
Always	4	34	53	54	70
No. of valid respondents	28	18	72	13	23

Recommendations of targeted N and accounting for nutrients in manure

Questionnaire results indicate that a large proportion of respondents already recommend best management practice for soil nutrients (Table 8.4, Fig. 8.3). Increasingly, advisers appear to be accounting for nutrients in manure (although this is not across the board) with 82% of all respondents stating that they had recommended this in the last 2 years. Wide distribution and use of MANNER and Managing Livestock Manures booklets support this high figure. A large proportion of all respondents (77%) also claimed they had recommended targeted N in the last 2 years. The proportion of agronomists recommending these practices is comparable with that of ADAS advisers. The high numbers of conservation advisers claiming to recommend targeted N is not coincident with levels of usage of RB209 or FACTS training amongst this community.

Agronomists and N recommendations: a focus on managing nutrients in manures

Despite widely reported use of tools amongst the agronomy community, analysis of interview data, which explores agronomist understanding and competence in recommending N and manure rates at the field level in more detail, suggests that the questionnaire data should be viewed more circumspectly. It appears that, rather than depending entirely on formal processes and technical tools when recommending N fertilizer applications, agronomists tend to rely on experience or 'user knowledge'. They use this knowledge as a reference point against which to compare outputs from tools and systems and then modify them to arrive at a recommendation to suit local circumstances, bringing together experience, and practical and local factors such as timing, soil type and stage of rotation. This method, however, becomes less effective when circumstances are unfamiliar and interviews revealed that adjusting nutrient budgets to allow for organic N supplied in manure applications is particularly challenging in this respect. Although agronomists have built up experience recommending N, they do not have the experience of high fertility situations where organic manures have been added, as one (Independent agronomist P) remarked:

> You could say that in known conditions with a regular arable rotation there is an intuitive level which we could establish with a reasonable level of confidence but in high fertility situations, this is where we come unstuck and struggle.

Estimating the nutrient value of manure requires some understanding of nutrient dynamics in the soil and being able to estimate amounts, and the nutrient content, of manure. These have proved unfamiliar skills for many agronomists, consequently achieving the required soil and crop nutrition is a complex process for which they often feel ill equipped as, one independent agronomist (P) said, 'Well who can cope with manures? You have to make a stab at it and estimate what's going on'.

These comments suggest that although agronomists are becoming FACTS-registered using tools to recommend efficient use of fertilizer N and consulting Managing Livestock Manure booklets, as indicated in the questionnaire data, they often refer to, and rely on, their own experience or user knowledge. This user knowledge however is challenged in unfamiliar situations where organic N is added in manure.

Discussion and conclusions

This analysis shows that members of the advisory community are generally knowledgeable about soil best management practice and appear to be observing (and have concern about) soil degradation, undertaking training, using guides and recommending soil best management practice to a relatively large extent. However further exploration of the data reveals a more complex picture, a focus on agronomists shows an apparent lack of hands-on experience, skills and involvement in cultivation practices, which is highly significant considering that cultivation decisions on-farm are central to maintaining good soil structure. Also it appears that although agronomists use tools to assist them in N recommendations, they also rely on their own experience and locally derived knowledge which fails in unfamiliar situations, such as when they have to account for nutrients supplied by manure. Ingram and Morris (2007) examine in more detail the nature of agronomists' local knowledge, derived from experience, and their scientific knowledge, derived from more formal sources, and discuss how these different forms interact.

Therefore, although results are generally positive with respect to the advisers' understanding of sustainable soil management, and the agronomists in particular, there are clearly some areas that need to be significantly enhanced and standardized to meet new policy challenges. Although formal training courses such as the new BASIS certificate in soil and water management are being offered to advisers (*Farmers Weekly*, 2004b), this research has revealed that advisers derive their knowledge and understanding from practical experience as well as from more formal sources. Opportunities for enhancing practical experience through interaction with competent practitioners using demonstrations and workshops should therefore be explored to complement training and the dissemination of publications. To some extent this is taking place already through NGOs and Defra funded initiatives but the need to enhance these and recognize the value of advisers learning through practical experience is clear.

In terms of the different categories, ADAS advisers demonstrate consistently high levels of training, use of guides, tools and recommendations suggesting that these advisers hold considerable knowledge and understanding about soil. This would be expected from an organization that is contracted by Defra to disseminate and research environmental protection. However, the ADAS advisers with expertise in soil management restrict their involvement to mass extension campaigns rather than to on-farm, one-to-one advice which will inevitably limit the impact of their soil management knowledge. Equally conservation advisers are limited in the extent of their impact, because, although they appear to be engaging more with practices beyond their core activities of habitat and wildlife conservation, and they provide one-to-one advice, they do not support everyday farming decisions. Agronomists however have a central role on the farm, they provide one-to-one regular arable advice which is integral to soil management. They are using guides, tools and recommending practices to nearly the same extent as ADAS advisers in many cases, which suggests that some agronomists at least are moving away from so called 'productivist' modes of thinking. Clearly they are recognizing the need to train in environmental practices both as part of Continuing Professional Development for FACTS/BASIS and to be able to meet more challenging farmers' requests. With enhancement of some areas identified in this study agronomists have the potential to make a positive contribution to the transition towards more sustainable soil management.

Notes

1. Questionnaires were emailed on behalf of the author through the FACTS organization to an unknown number and sector of the membership. FACTS provides a national training syllabus and accreditation for arable advisers and certification ensures a certain standard of knowledge of fertilizer (inorganic and organic) application and of the Soil Code.
2. This term agronomist is used here synonymously with crop consultant.

3. Using correlations, three categories of recommendation relationships between the practices were identified. Significant (at 0.01) correlations (2 tailed Pearson) were found between numbers of respondents recommending the cultivation practices of carefully timed cultivation, low compaction machinery, earlier autumn sowing and minimum tillage; between numbers recommending fertilizer practices of targeted N and N value of manure; and between numbers recommending the anti-erosion measures of buffer strips and contour ploughing.

References

ADAS, IGER and SRI (2000a) *Managing Livestock Manures. Booklet 1: Making Better Use of Livestock Manures on Arable Land.* ADAS, Gleadthorpe Research Centre, Mansfield, UK.

ADAS, IGER and SRI (2000b) *Managing Livestock Manures. Booklet 2: Making Better Use of Livestock Manures on Grassland.* ADAS, Gleadthorpe Research Centre, Mansfield, UK.

Angell, B., Chalmers, A., Flint, C. and Francis, J. (1997) *Agriculture and the Rural Economy Information and Advice Needs.* Report by ADAS to MAFF, London, UK.

Archer, J. (2001) *Review of Current and Recent Knowledge Transfer Activities in Relation to Environmental Protection in Agriculture.* Report to Rural and Marine Environment Division, MAFF, London, UK.

Burton, M., Rigby, D. and Young, T. (1999) Analysis of the determinants of adoption of organic horticultural techniques in the UK. *Journal of Agricultural Economics* 50(1) 47-63.

Cooper, N.K. (1999) *Street Level Bureaucrats and Agri-Environmental Schemes: The Case of the FRCA Project Officer Implementing ESA Schemes in England and Wales.* Unpublished PhD Thesis, King's College, London, UK.

Coughenour, C.M. (2003) Innovating conservation agriculture: the case of no-till cropping. *Rural Sociology* 68(2) 278-304.

Cox, G., Lowe, P. and Winter, M. (1990) *The Voluntary Principle in Conservation: A Study of the Farming and Wildlife Advisory Group.* Packard, Chichester, UK.

Curry, N. (1997) Providing new environmental skills for British Farmers. *Journal of Environmental Management* 50, 211-222.

Dampney, P., Winter, W. and Jones, D. (2001) *Communication Methods to Persuade Agricultural Land Managers to Adopt Practices that will Benefit Environmental Protection and Conservation Management (AgriComms).* Report by ADAS to Defra, London, UK.

Dancey, R.J. (1993) The evolution of agricultural extension in England and Wales. *Journal of Agricultural Economics* 44(3) 375-393.

Dauven, A., Crabb, J. and Johnson, P. (1998) *Nitrate Vulnerable Zones - Baseline Survey.* Report to RMED, MAFF, London, UK.

Dauven, A. and Crabb, J. (1999) *Survey of Sugar Beet Growers' Information Needs.* Report to Sugar Beet Research and Education Fund, MAFF, London, UK.

Defra (2004a) *The First Soil Action Plan for England 2004-06.* Defra, London, UK.

Defra (2004b) *Developing Measures to Promote Catchment-Sensitive Farming.* A joint Defra-HM Treasury Consultation. Defra, London, UK.

Defra (2004c) *Learning, Skills and Knowledge Review Final Report.* Defra, London, UK.

Defra (2005) *Controlling Soil Erosion. A Manual for the Assessment and Management of Agricultural Land at Risk of Water Erosion in Lowland England.* Defra, London, UK.

Defra (2006) *Cross Compliance Guidance for Soil Management.* Defra, London, UK.

ECOTEC (2000) *Economic Evaluation of Free Advice Programmes on Conservation, Pollution Abatement and Conversion to Organic Farming.* Report C1752 to MAFF, London, UK.

Eldon, J. (1988) Agricultural change, conservation, and the role of advisers. *ECOS* 9(4) 15-20.

Environment Agency (2001) *Best Farming Practices: Profiting from a Good Environment.* R&D Publication 23. Environment Agency, Bristol, UK.

Environment Agency (2004a) *Soil, the Hidden Resource. Towards an Environment Agency Strategy for Soil Protection, Management and Restoration. A Consultation Document.* Environment Agency, Bristol, UK.
Environment Agency (2004b) *The State of Soils in England and Wales.* Environment Agency, Bristol, UK.
Farmers Weekly (2004a) Soil help needed to keep SFP. *Farmers Weekly,* 13 May 2004. [Accessed June 2004] Available from *Farmers Weekly Interactive,* http://www.fwi.co.uk/live/
Farmers Weekly (2004b) Certificate to suss soil. *Farmers Weekly,* 15 June 2004. [Accessed June 2004] Available from *Farmers Weekly Interactive,* http://www.fwi.co.uk/live/
Farmers Weekly (2006) Agronomists go green on advice. *Farmers Weekly*, 13 Jan 2006. [Accessed Jan 2006] Available from *Farmers Weekly Interactive,* http://www.fwi.co.uk/live/
Fearne, A. (1991) Agricultural information: a farmer's point of view. In: Kuiper, D. and Roling, N.G. (eds) *European Seminar on Knowledge Management and Information Technology.* Department of Extension Science, University of Wageningen, The Netherlands.
Garforth, C., Angell, B. and Archer, J. (2003) Fragmentation or creative diversity? Options in the provision of land management advisory services. *Land Use Policy* 20(4) 323-333.
Gasson, R. and Hill, B. (1996) *Socio-economic Determinants of the Level and Rate of On-farm Innovation.* Vol.1. A review of literature. Vol.2. Mechanisms of technology transfer in agriculture and horticulture. Wye College, University of London, UK.
Ingram, J. (2008) Are farmers in England equipped to meet the knowledge challenge of sustainable soil management? An analysis of farmer and advisor views. *Journal of Environmental Management* 86(1) 214-228.
Ingram, J. and Morris, C. (2007) The knowledge challenge within the transition towards sustainable soil management: an analysis of agricultural advisors in England. *Land Use Policy* 24, 100-117.
Jones, G.E., Rolls, M.J. and Tranter, R.B. (1987) *Information Management in Agriculture.* British Library Review and Report 5931. British Library, London, UK.
JT Research Agribus (2003) *Farmers Weekly Group.* [Accessed 1 Jan 2004] Available from http://www.farmersweeklygroup.co.uk
Juntti, M. and Potter, C. (2002) Interpreting and reinterpreting agri-environmental policy: communication, trust and knowledge in the implementation process. *Sociologia Ruralis* 42(3) 215-232.
MAFF (1998) *Code of Good Agricultural Practice for the Protection of Soil.* MAFF Publications, London, UK.
MAFF (1999a) *Controlling Soil Erosion by Water: a Manual for the Assessment of Agricultural Land at Risk of Water Erosion on Lowland England.* MAFF Publications, London, UK.
MAFF (1999b) *Controlling Soil Erosion by Water: a Field Guide for an Erosion Risk Assessment for Farmers and Consultants.* MAFF Publications, London, UK.
MAFF (2000) *Fertiliser Recommendations for Agricultural and Horticultural Crops. RB209.* Seventh edition. The Stationery Office, London, UK.
Marshall, E.J.P. (2002) *Environmental Information for Agronomists 2002: Needs and Provision.* A report by the Farmed Environment Company for Crop Protection Association UK Ltd and UKASTA, UK.
Morris, C. and Winter, M. (1999) Integrated farming systems: the third way for European agriculture? *Land Use Policy* 16, 193-205.
OECD (2001) *Adoption of Technologies for Sustainable Farming Systems.* Workshop Proceedings. The Agricultural University, Wageningen, The Netherlands.
Park, J., Farmer, D.P., Bailey, A.P., Keatinge, J.H.H., Rehman, T. and Tranter, R.B. (1997) Integrated arable farming systems and their potential uptake in the UK. *Farm Management* 9(10) 483-494.
Roling, N.G. and Wagemaker, M.A.E (eds) (2000) *Facilitating Sustainable Agriculture.* Cambridge University Press, Cambridge, UK.
Royal Commission on Environmental Pollution (RCEP) (1996) *Sustainable Use of Soil.* RCEP Nineteenth Report. HMSO, London, UK.

Tebrugge, F. and Bohrnsen, A. (2001) Farmers and experts opinion on no-tillage in West Europe and Nebraska. In: Garcia-Torres, L., Benites, J. and Martinez-Vilela, A. (eds) *Conservation Agriculture. A World-wide Challenge.* Vol. 1, XUL Press, Cordoba, Spain, 61-71.

UK Soil Management Initiative (SMI) (2005) *Visual Soil Assessment.* UK SMI, Chester, UK.

Wilson, G. (2001) From productivism to post productivism and back again? Exploring the (un)changed natural and mental landscapes of European agriculture. *Transactions of the Institute of British Geographers* 26(1) 77-102.

Winter, M. (1995) *Networks of Knowledge: A Review of Environmental Advice, Training, Education and Research for the Agricultural Community in the UK.* Report to WWF. Countryside & Community Research Unit, University of Gloucestershire, Cheltenham, UK.

Winter, M. (1996) Landwise or land foolish? Free conservation advice for farmers in the wider English countryside. *Landscape Research* 21(3) 243-263.

Winter, M. (1997) New policies and new skills: agricultural change and technology transfer. *Sociologia Ruralis* 37(3) 363-381.

Winter, M., Mills, J., Lobley, M. and Winter, H. (2001) *Knowledge for Sustainable Agriculture.* Report to WWF. Countryside & Community Research Unit, University of Gloucestershire, Cheltenham, UK.

Chapter 9

Combining scientific and lay knowledges: participatory approaches to research in organic farming

F. Lyon and F. Harris

Introduction

While agricultural research has led to dramatic technological change since the late 1950s, emphasis has been on maximizing production for high input agriculture in high potential areas. There has been less research on complex and diverse farming systems, characteristics typical of many farming systems, in particular organic farms. Much of the existing research has prioritized the roles of scientists and on station trials while users of the research, particularly farmers, are perceived as passive recipients of technology, what Biggs (1990) refers to as the central source model. However, empirical studies show that there are multiple sources of innovation, including agri-businesses, universities, farming organizations and farmers themselves. Farms are also small businesses and research on the innovation processes of such firms shows they have the propensity to research and learn (Smallbone *et al.*, 2003). This suggests that formal scientific research would benefit from closer integration with the users of technology, not just in setting the research agenda but also in carrying out the research.

This chapter explores the issues of different types of knowledge and how they can be used to complement each other in organic farming research. Two specific issues are addressed in the context of plant breeding research:

- What types of research and informal learning are carried out by farmers?
- How can farmers' own research add to and complement more traditional scientific methods of research?

Since the late 1980s, there has been a decline in public sector funding for agricultural research internationally, privatization of public research centres and a dramatic increase in private sector investment in research. Public sector research has been focused on pure or basic agricultural research and those areas that are considered to be 'public goods' such as environmental sustainability (Heisey *et al.*, 2002). Private sector research, from plant breeding and agrochemical companies for example, has been customer focused but has concentrated on the needs of larger 'conventional farmers' and the major buyers of the products, most notably supermarkets.

Smaller agricultural businesses, such as organic farmers and the buyers of their products, represent a smaller market which is not considered to be as profitable an investment for agricultural research compared to conventional plant breeding. This is due, first, to the absolute size of the market and amount of seed required, and second, to the small number of dispersed businesses which represent higher marketing costs (Heisey *et al.*, 2002). Furthermore, the close relationship between plant breeding and producing agro-chemicals (in many cases by the same company) results in less interest in research for organic agriculture.

Organic agriculture emerged through the interest and determination of farmers committed to alternative farming methods. Thus its development came about through participatory research and development, in which the main participants were the farmers themselves, acting mainly as farmers but also as consumers and citizens. Due to the lack of outside support, farmers had to work together, exchanging ideas and experiences to enhance their knowledge and expertise. By developing in this broad but pragmatic fashion, organic farmers have shown that there are alternatives to the traditional scientific model of agricultural learning and extension.

Defining experimentation and research

There is considerable debate in the literature concerning the differences between what farmers do and the generation of 'modern scientific knowledge' (Kloppenberg, 1991; Molnar *et al.*, 1992; Scoones and Thompson, 1994). To summarize the debates, 'scientific knowledge' is seen to relate to concepts, practices and technology that are based on tried and tested theories leading to universally applicable results, or results related to a specific set of conditions. Knowledge is generated through rigorous procedures that attempt to control variables in order to produce quantitative results for statistical analysis. Robust results can be used to convince others, make confident recommendations and can also be extrapolated to different contexts. At the same time, farmers, grain traders and food processors are doing their own research. This research may be carried out in a less rigorous manner, often through observing their own experiences, and they may not refer to it as research or experimentation, but it is an important way of learning and generating 'anecdotal' evidence. The quotation below is an example of how one farmer has reflected on these ways of learning, although his sentiments are frequently aired by others:

> I am a researcher of sorts. Like all farmers, I constantly 'experiment' with things like seed rates (not always deliberately), timing of cultivations, leave a strip there and see what happens, do that later and see what happens, feed this now and see what happens etc., etc. The linking of anecdotal research to scientific or pure research could prove valuable and running the two in parallel makes real sense. All knowledge comes directly or indirectly from experience and reflection on that experience. That is something we should not forget or ignore (Wilson, 2002: 2).

Farmers' research is closely linked to coping with a changing environment, and informal experimentation (planned and unplanned) is a major way of learning. They tend to adapt rather than adopt new innovations in order to fit their system, and this becomes increasingly important when research programmes do not have the capacity to 'iron out the levels of uncertainty in recommendations in the more diverse and complex farming systems' (Ashby, 1987: 239).

The process of farmers' experimentation may not appear as the testing of structured hypotheses so a wider understanding of experimentation is needed that goes beyond positivist definitions (Scoones and Thompson, 1994). Okali *et al.* (1994) propose a broader approach with two essential aspects: first the creation of treatments or initial observation of conditions; and second observation and monitoring of results or effects. Based on these two aspects, methodologies of farmer experimentation can be divided into:

- Hypothesis testing for solving problems, adapting technology or for curiosity (Rhoades and Bebbington, 1991). This may be similar to conventional scientific research;
- Experiential research and 'accidental' experiments. Experiments can come about when a farmer is forced to try something new or when a chance event such as weather or a mistake allows them to make an observation (Dyke, 1974: 4; Richards, 1994; Stolzenbach, 1992: 159; Lyon, 1996).

Farmers' research and knowledge generation is based on more holistic and open system views, grounded in a particular local context. Measuring, observing and evaluating the results may not be done as rigorously as in conventional scientific research. Farmers may have a different concept of 'check' to researchers, with validation coming from 'inter- instead of [as well as] intra-farm replication' (Farrington and Martin, 1987: 62), comparison to previous years and peer critique (Richards, 1994). Lyon (1996) found that farmers visit farms of their friends and neighbours informally and discuss ideas while 'talking shop' down the pub. Farmers assess the experiment on a range of interacting criteria as part of their farming system. The complexity of interaction and multiple criteria means farmers often have to rely on a 'gut feeling' (Lyon, 1996).

As mentioned earlier, innovation may not be seen as research or experimentation but as the way people learn. In this way innovation is part of informal organizational learning (Porter, 1990; Smallbone *et al.*, 2003), and part of what Bessant *et al.* (1994) refer to as 'continuous improvement'. In the experiential approach to research and learning, farmers evaluate through what Schon (1983) calls 'reflection-in-action' either during the action ('thinking on your feet', 'learning by doing' or 'keeping your wits about you') or after (either immediately or over the course of several months). The evaluation is a continuous process through examining the direct impact of the operation throughout the season rather than only evaluating the output. It can be done consciously, such as a 'post-mortem' on a situation, or subconsciously and form part of the farmers' tacit knowledge.

Tacit knowledge is hard to define as it is based on the premise that 'we can know more than we can tell' (Polanyi, 1966: 4) and it is difficult to write down or formalize. Ambrosini and Bowman (2001) characterize it as mental models that individuals can follow in certain situations. Nonaka (1991: 98) compares it to the skills referred to as a 'know how': 'deeply rooted in an action and in an individual's commitment to a specific context - a craft or a profession, a particular technology or product market, or the activities of a work group or team'.

Ambrosini and Bowman (2001) introduce the concept of degrees of tacitness ranging from explicit skills, to tacit skills that can be articulated, to tacit skills that can be imperfectly articulated, to deeply ingrained tacit skills. The first two categories can be built up through intentional learning with the knowledge being internalized later, while the latter two categories may arise though a subconscious process of reflecting on behaviour.

Methodology for assessing existing systems of innovation and interaction

This chapter is based on research undertaken in 2003-4 which involved semi-structured interviews with a sample of 30 farmers and other agri-businesses in East Anglia and South West England. These were selected purposely to ensure a cross section of types of farmers and other businesses. Farmers were selected from those working with Elm Farm Research Centre (EFRC), conventional farmers working with private seed companies, and other farmers who may not be so well networked. Interviews were also carried out with seed producers/dressers/sellers, grain buyers/millers, and other end users. In addition seven scientists were interviewed, including people working for organic research stations, public sector funded research stations and private sector research stations.

The semi-structured interviews followed a check list of questions, while allowing the interviewee to explore some issues in detail, such as the processes by which farmers do their own research. Interviews were taped and partially transcribed. A particular challenge was to identify and gather data on types of tacit research and innovation. This requires particular care in the conduct and analysis of interviews. Interviewees were asked to describe the reasons why things happen and to talk about particular instances (critical incident analysis - Chell, 1998). Tacit knowledge can be identified when people have to think for some time, laugh at a question or make statements such as ‘oh yes, that’s right’ or ‘Aha, I hadn’t realized that’ or ‘I’ve never really thought about that’ (Ambrosini and Bowman, 2001: 820). The use of metaphors, similes and analogies were also noted as these can be ways of giving tacit knowledge a voice when explicit language cannot explain it (Ambrosini and Bowman, 2001).

Types of research and experimentation involving farmers and other businesses

All farmers interviewed were found to be doing some form of experiment or research or trying new activities. This was expressed by one farmer as follows: ‘Usually each year we try two or three things, just something a little bit different to just try and sort of push the boundary a bit and see whether we can come up with something’ (Interview 4). A wide range of subjects for experimentation was discovered and every farm was found to be comparing varieties grown between fields, in previous years, or on other farms. Similarly, agronomy advisors were continually making comparisons between varieties grown on each farm and between farms. Some farmers could be stimulated to experiment after learning of the results of existing research, either by scientists or by other farmers. However, others were more risk averse and unwilling to try varieties that were not well established, as the following quotation highlights:

> We grow 40 acres of corn, say we put in 10 acres of a new variety and it absolutely fails, you've lost 25% of your corn. Whereas with the old variety there's less risk of failure because at least you know it's performed before at your farm. But then somebody's going to test it and give it a try (Interview 6).

Seed dealers were found to be carrying out trials in partnership with seed growing farmers, with an agreement that if the trial failed there would be compensation. Seed dealers also carried out informal trials through comparing different varieties on different farms and noting growing conditions. A range of different innovations and approaches have been tested in relation to the marketing of cereals. While this was not an issue raised by many farmer interviewees, it was a primary concern of grain buyers. These enterprises were particularly concerned with a lack of farmers' interest in meeting market demand in terms of crops produced and quality or consistency of products:

> The trouble is that farmers tend not to be very market-driven … it's fascinating to grow but if you can't sell it … Organic farmers are tending to grow crops that nicely fit organic rotations or are easy to produce (Interview 16).
>
> It's the arrogance of it. There are still a lot of organic ones that seem to think, well we've grown it, and you've got to take it (Interview 17).

Design of farmers' own experimentation

The study also explored how farmers carried out experiments. The design of farmers' trials can be a conscious decision at the beginning of the farming season or it can take place during the season as farmers react to specific problems, growing conditions or accidental treatments. For specific treatments farmers commonly leave part of the field untreated to see if there is any effect. For example, an unusually research-minded farmer chose to run trials of 16 winter wheat varieties, and 35 varieties of other arable crops:

> I don't believe in following the pack. We have to go down that route that best serves us and if others benefit then that is fine … we are very flexible. It is science but we are not governed by the same things. For example I had not planned to top dress with compost, but the crop looked sad and so I thought we could try that because we were going to trial compost next year (Interview 1).

This case shows how farmers can combine a more scientific approach to experimentation with the practice of farming, which means that variables have to change through the season. In many other cases farmers did not have clear hypotheses at the start and the nature of the trial became apparent partway through the season as farmers reacted to particular events or accidents. Examples include turning a cereal crop into whole crop silage, and mixing peas with triticale. The design of trials may come about by accidents occurring through the farming season or having to respond to particular problem. For example, one farmer was left with an accidental mixture of crops in a field due to a mistake during sowing but as the crops had grown very well he repeated the mixture in future years. Another recounted his trialing process as follows:

> I have tried different drilling times, but often by accident … just by circumstance. I found that the later sown crops out-yielded others so I am convinced. The temptation is to put it in early, especially when there is a mild autumn. But now I've tried to go later (Interview 2).

Farmers' analysis of their trial results

Farmers evaluate according to a very wide range of criteria, although yield is of most importance to them. This is assessed by a variety of means, including weighing in grain dryers, scales in combines, time taken to fill a combine, or the number of trailer loads (especially for forage crops). Other information collected includes straw height, levels of competition with weeds, and milling quality. This process is illustrated by one farmer's account of his evaluation of a blend of wheats - Maris Widgeon and Petchworth - grown in the previous year. He assessed the mixture through the season noticing that Maris Widgeon's tall straw resulted in competition for weeds on a bit of land that was previously 'very dirty'. Compared to his other crops, the mixture gave the best yield (he used a dryer that can also weigh) and the best quality for milling, seen from the agber count. He also assessed the crop visually in terms of weed growth and visitor expectations:

> It has to look good. I do not want it overflowing with weeds, because we have to pay for that further down the line and I want it to look good for the people coming around (Interview 2).

Taking measurements can be time consuming and inconvenient. A sense of these difficulties is conveyed in the following farmer's account of the process involved in assessing yields:

> If you want to treat half of an organic wheat field and then assess the yields at the end of that, it's just more trouble. Our combine doesn't weigh yield as it goes along as some of the new combines do. So what we have to do is weigh it as it goes through our dresser (Interview 9).

Yield could also be assessed by a general feeling of how the combine is coping with the crop. Tractor drivers and farm workers can therefore play an important role in assessing experiments.

Advisors play a key role in setting up trials and making cross-farm comparisons. Many recently converted organic farms have retained their conventional agronomists as they have a long standing knowledge of each field in terms of soil type and weed pressures. Likewise informal networks of farmers and farm walks are important sources of comparisons of crops with different treatments. One farmer recounted how 'I share ideas with other farmers and … our advisor comes around every six months and we chew the fat and throw around ideas' (Interview 2). Another explained in more detail:

> I might be thinking about doing something that can be completely off the wall and I can say 'look I've been thinking about doing this, do you know

> anybody who might be doing this, or do you think this is completely daft?' It's quite interesting to hear that you're not the only one doing it or that somebody else is trying to resolve it another way. With the new things we are doing, we've talked about it and that perhaps we ought to be trying to quantify it more than we have been because we've probably been down the subjective route so far and perhaps we'd like to try to be a bit more objective about it (Interview 19).

The most common form of analysis of experiments is through comparisons to previous years, crops on adjacent land or crops sown at the same time. The farmers displayed a detailed knowledge of some of the factors that might have caused the difference, based on years of experience of working the land and building up local knowledge of how each field might respond to different conditions. However, farmers' knowledge can be limited or even wrong. Through reducing the possible variables and excluding the treatments and conditions that were similar, farmers can start to attribute cause and effect. One farmer described the process as follows:

> If there is a big difference between two fields, the first thing I suppose would be to look at what we had done to it. So in term of drilling dates, seed rates, cultivations, seed bed preparation, more weeding on one, some muck on or not … All those sorts of things, try and, ok well we did that on that field and didn't do it on that one or whichever. Then look at obvious things would be soil type, soil analysis … timeliness of harvest. Yeah, I guess those would be the obvious things and then if that throws anything up it would dictate whether I thought oh gosh there's a difference here or maybe it's just the way it is (Interview 6).

Implicit assessments

The actual assessment of crops may not be done formally. One farmer stated 'it just doesn't feel like it is research it's just … I mean over the years we might see the difference' (Interview 9). Other interviewees denied doing any experimentation until part way through the interview and they had reflected on why they were carrying out their operations as they were. Many of the innovations concerning machinery were referred to as 'tinkering' and involved applying tacit knowledge held by farm workers or technicians, rather than farm managers. Such assessments could be carried out almost subconsciously and may not be explicitly acknowledged until asked the question, as shown in following exchange:

> Interviewer: So how will you know that it is working?
> Farmer: That's a very good question … I will compare … I guess it'll be just a visual monitoring, the important bit will be yield, the important bit will be whether one comes down with a load of disease that the other one didn't, how does it germinate, all those sort of things. I don't think I'll be doing it particularly scientifically necessarily, it'll just be an overall impression that, that worked or it didn't work. …
> We've just changed to use a contractor who has two drills and we have one too. We also tried a demonstration drill for about forty hectares. How do you know which is better? I guess it'll be to do with seed placement and coverage

> … number of cultivations … germination … you just sort of log these things away I think as you go around but then … it might then prompt you to actually go back and try to be a little bit more scientific. But I guess because it is in the back of your mind it's just for your own personal use, so you don't tend to record it all and say 'oh right that's how to justify XYZ'. It's more of a sort of, I think … it's driven by something you notice yourself and then you think … ah maybe I'd better try and put a bit more of a handle on it (Interview 4).

Interactions with scientists

Scientific researchers and farmers have differing objectives and agendas that need to be better understood in order to promote cooperation and participation. To a large extent scientific knowledge is disseminated to farmers through intermediaries, such as technical advisers or agronomists. These individuals can act as bridges between the differing cultures.

Within organic agriculture there is very little research in the commercial sector as seed breeding companies do not consider the size of the organic seed market to be viable for a specific organic breeding programme. However, some commercial breeders are promoting conventional varieties that are well suited to organic systems, and have an interest in organic farming because of concerns over disease resistance, and for public relations. This tends to be the smaller organizations that are not commercially connected to agrochemical sales and are content to capture small proportions of the market with varieties that meet specific needs of farmers or users.

Where farmers in our survey had been involved with researchers, there appeared to be considerable misunderstanding present with both sides becoming frustrated. Farmers could be highly critical of research projects they felt had not been relevant:

> We had a research project and we sat around the table, with professors and PhDs and when I mentioned anything off the beaten track they just … they wanted to do research on their own topics. They did not understand organic systems so it was a huge waste. Some of the stuff was really bad. There was lots of money but I had the feeling that this was a jolly. I was a bit disillusioned. Then at the end they said they had run out of money, so could not write it up. No one else could say that and get away with it. If it was a business, you would not run it that way. I used to bleat mildly, but now in retrospect I would make more fuss. Then I was sitting there with all these high powered people. They're not like X. He is an academic with wisdom - he was saying that we have to have the anecdotal, with the scientific. But then this view got filtered down to the lower academics, and they went on their own way … they were very airy fairy. In this case it was hijacked (Interview 2).

There was also some concern over the reliability of results coming from commercial trials:

> Some scientists are very bigoted and use the research to find what they want to achieve. Some scientists do what they say and some companies are better

> than others; independent trials are better as long as they are done by people with no axe to grind (Interview 10).
> Sometimes it's … how can I say … done by big companies and they do the research and it doesn't matter where they do the research their product is always better than everybody else's. Because basically if there's five in a group and theirs beats three, they'll show you the ones they beat, they won't show you the one that actually beat them (Interview 6).

Research coming out of universities and research stations is often criticized by farmers and other businesses in the food chain for its lack of validity in the 'real-world'. A common complaint is that scientists are not working in the commercial context and lack knowledge about the organic food industry making the results of their research less valid. In part this issue is due to the reductionist methods (such as small plot trials and the rigour associated with the need for statistics) that allow scientists to do more basic research. Research institutes increasingly have to follow this agenda as well in order to demonstrate the quality of their work to public sector funding organizations. One researcher stated:

> What farmers want is messages, they want very simple, very clear messages. The problem with these refereed papers … they've got things like, 'this maybe, probably means that', it's all sort of 'ifs' and 'buts' because you can never be 100% sure. But farmers don't want that, they want to hear that we've done it and we believe this is the way that this will do well (Interview 22).

However, the research scientists that were interviewed from universities and research centres were aware of these criticisms and stated that they are under increasing pressure to publish papers in journals targeted at other academics and which are therefore written in a way that may not be accessible to farmers. Particular challenges were faced by those trying to publish work that was not based on replicated trials, such as work on systems:

> Systems work does not have replications and in the science world this becomes very difficult. They complain that we are only looking at one farm and the replication is only over several years. We hawked a paper around several journals but the editors said they could not accept it in the end because there were no replicates (Interview 24).

A lack of satisfaction with the existing forms of disseminating results to farmers was voiced by four of the interviewees and was the conclusion of a recent meeting of farmers and scientists:

> There is quite a lot of knowledge and information out there and sometimes it's not necessarily about doing experiments in their own right, it's more about collating that information and actually making it available (Interview 4).

The process of dissemination can also stimulate more farmer experimentation and encourage farmers to talk to other farmers about their experiences of dealing with those issues. One farmer was keen to establish new, more interactive research exchanges which would facilitate this:

> Let's have some new ways where farmers can poke around with crops as farmers do and say, 'bloody hell you might like it mate I don't' and 'you ain't seen my farm', 'what about the yield', you know all of that (Interview 16).

Examples of farmer-scientist collaborations

Our small survey identified 20 cases of farmer-scientist collaboration. The different forms of interaction ranged from those where the scientists controlled the design and analysis to those where farmers had more control.

Researcher managed and implemented

The most commonly reported and most easily observable forms of interaction are on-farm and on-station trials that are designed and analysed by researchers. The types of trials reported to us were predominantly varietal trials. While the objective of such trials is to generate statistical information that has relevance to other locations, farmers can gain information from observing them and seeing what is relevant for their particular farm. Farmers and other stakeholders can also interact in researcher managed trials by being on steering groups and advising on selection of research questions and types of treatment to be assessed. Typically researchers decide the varieties, sow and harvest, and farmers are consulted on appropriate trial locations.

Three of the farmers interviewed had had Elm Farm Research Centre trials on their land. They stated that they benefited from access to the results which were relevant to the fields that they use. Similarly farmers and other related businesses, such as seed dealers, are invited onto research stations or 'experimental farms' belonging to seed breeding companies or independent research stations. This gives them the opportunity to pick up new information, gather stimulating ideas for their own experimentation and, in return, provide the researchers with feedback on their existing varieties and ideas for future research. Feed manufacturers and millers can also be involved in some aspects of variety development in order to identify specific needs in terms of nutrition and identify varieties for specific purposes.

Researcher managed and farmer implemented

A number of on-farm research projects had been designed by researchers but carried out by farmers on their own land as part of their normal farming practice. Examples include variety and weed control trials, bird surveys and schemes to prevent pollution of water sources. These types of research are either designed as trials or the researcher involvement may involve monitoring what farmers are doing and the impact this has. Such studies may have minimal farmer involvement (such as surveys of the bird populations on organic and conventional farms) although they attempt to encourage greater farmer participation in analysing the results.

Getting farmers involved in research was, however, noted as a challenge by researchers. A team member of a project looking at the economics of organic farming in different locations outlined the difficulties faced:

> We wanted more participation from them, we wanted more of their time and we realized we had to pay them. Certainly the ones we've got now enjoy it; it's also a chance for them to meet with a few colleagues they value. There is no problem with getting conversation going. It's stopping them that's the problem. You can't get them away at the end of that day because they all know each other well and we have very, very good discussions (Interview 12).

Likewise a member of staff from a university based research centre also outlined the careful procedures involved to help secure the on-going participation of farmers:

> The Centre carries out on farm field trials, we tend to work with model farmers, those defined as being more forward looking and risk takers. When you first start working together you need to adapt to them and they need to adapt to you. You don't begin with a risky hypothesis the first time because if it is risky it might fail and farmers will not wish to work you again. It is important to develop a relationship of trust with the farmers so that they would be willing to continue on to new projects, rather than do one trial and then to give up (Interview 15).

Farmer managed with researcher involvement

The third type of farmer-researcher interaction relates closely to farmers' own research but involves researchers feeding ideas to farmers. An example of this is seed breeding companies providing seed to a farmer who is carrying out a range of varietal trials. Commercial scientists were also found to be suggesting varieties to a seed dealer who in turn asked farmers to try them out. Likewise our survey revealed that researchers were involved in advising farmers informally with regard to specific questions concerning soil quality and nutrient levels.

Conclusions

This study of farmers and businesses found that almost all were doing some form of experimentation in order to adapt new innovations to their specific farming ecosystems. While some farmers (and other agribusinesses) undertake experiments that follow scientific methods, others make holistic assessments using multiple criteria in a way that is not possible in conventional scientific, reductionist research. Farmers may set hypotheses explicitly before starting the experiment or they may use gut feelings and be experimenting without acknowledging it. The scientific methods, the holistic and the implicit approaches all have a contribution to make to agricultural research.

Many farmers are critical of existing public sector funded research for selecting irrelevant topics, and using small plot trials that are very different to the commercial context of farming. Scientists recognize this issue but reported that they are under pressure to publish in academic journals that demand the rigor derived from replicated plots trials. Farmers were also sceptical of results of research that had been funded by the private sector, particularly those developing technology and carrying out plant breeding.

Four types of farmer-scientist interactions were identified: scientist managed research on farmers' land; farmers invited onto research stations; scientific monitoring of farmers' own operations; and farmers' own research with researchers involved in providing ideas. There are challenges for scientists working with farmers as the latter may not pursue scientific rigor and may change treatments during an experiment. Where rigorous detailed statistics are required a more reductionist approach with greater researcher control may be needed. This can be complemented by understanding farmers' own research as well. At present farmers' knowledge is not built into research programmes although there is a range of key stakeholders who can encourage farmers and agri-businesses to experiment with new ideas. Participatory research should work with these advisors, input sellers and crop buyers.

References

Ambrosini, V. and Bowman, C. (2001) Tacit knowledge: some suggestions for operationalization. *Journal of Management Studies* 38(6) 811-829.

Ashby, J.A. (1987) The effects of different types of farmer participation on the management of on-farm trials. *Agricultural Administration and Extension* 25, 235-252.

Bessant, J., Caffyn, S., Gilbert, J., Harding, R. and Webb, S. (1994) Rediscovering continuous improvement. *Technovation* 14(1) 17-29.

Biggs, S. (1990) A multiple source of innovation model of agricultural research and technology promotion. *World Development* 18(11) 1481-1499.

Chell, E. (1998) Critical incident technique. In: Symon, G. and Cassell, C. (eds) *Qualitative Methods and Analysis in Organisational Research - a Practical Guide.* Sage, London, UK, 51-72.

Dyke, G.V. (1974) *Comparative Experiments with Field Crops*. Butterworths, London, UK.

Farrington, J. and Martin, A. (1987) *Farmer Participatory Research; A Review of Concepts and Practices*. Agricultural Research and Extension Network Discussion Paper 19. ODI, London, UK.

Heisey, P.W., Srinivasan, C.S. and Thirtle, C. (2002) Privatization of plant breeding in industrialized countries: causes, consequences and the public sector response. In: Byerlee, D. and Echeverria, R. (eds) *Agricultural Research Policy in an Era of Privatization*. CAB International, Wallingford, UK, 177-198.

Kloppenberg, J. (1991) Social theory and the de/reconstruction of agricultural science: local knowledge for an alternative agriculture. *Rural Sociology* 56(4) 519-548.

Lyon, F. (1996) How farmers research and learn: the case of arable farmers in East Anglia, UK. *Journal of Agriculture and Human Values* 13(4) 39-47.

Molnar, J.J., Duffy, P.A., Cummins, K.A. and Van Santen, E. (1992) Agricultural science and agricultural counterculture: paradigms in search of a future. *Rural Sociology* 57(1) 83-91.

Nonaka, I. (1991) The knowledge creating company. *Harvard Business Review* 69(6) 96-104.

Okali, C., Sumberg, J. and Farrington, J. (1994) *Farmer Participatory Research: Rhetoric and Reality*. Intermediate Technology Publications, London, UK.

Polanyi, K. (1966) *The Tacit Dimension.* Doubleday, New York, USA.

Porter, M.E. (1990) *Competitive Advantage of Nations.* The Free Press, New York, USA.

Rhoades, R. and Bebbington, A. (1991) Farmers as experimenters. In: Haverkort, B., Van de Kamp, J. and Waters-Bayer, A. (eds) *Joining Farmers' Experiments*. Intermediate Technology Publications, London, UK, 251-255.

Richards, P. (1994) Rural development and local knowledge; the case of rice in Sierra Leone. In: Scoones, I. and Thompson, J. (eds) *Beyond Farmer First: Rural People's Knowledge,*

Agricultural Research and Extension Practice. Intermediate Technology Publications, London, UK, 165-169.

Schon, D. (1983) *The Reflexive Practitioner. How Professionals Think in Action*. Basic Books, New York, USA.

Scoones, I. and Thompson, J. (eds) (1994) *Beyond Farmer First: Rural People's Knowledge, Agricultural Research and Extension Practice*. Intermediate Technology Publications, London, UK.

Smallbone, D., North, D. and Vickers, I. (2003) The role and characteristics of SMEs in innovation. In: Asheim, B.T., Isaksen, A., Nauwelaers, C. and Todtling, F. (eds) *Regional Innovation Policy for Small-Medium Enterprises*. Edward Elgar, Cheltenham, UK, 3-20.

Stolzenbach, A. (1992) Learning by improvisation: the logic of farmers experimentation in Mali. Paper prepared for IIED/IDS *Beyond Farmer First: Rural Peoples' Knowledge, Agricultural Research and Extension Practice Workshop*, 27-29 Oct 1992. Institute of Development Studies, University of Sussex, UK.

Wilson, D. (2002) Farmers and Organic Research. *Proceedings of the UK Organic Research 2002 Conference*. Aberystwyth, Wales. [Accessed July 2007] Available from http://www.organic.aber.ac.uk/library/Farmers%20and%20organic%20research.pdf

Chapter 10

Sustainable foodscapes? Examining consumer attitudes and practices towards food and farming in Washington State, USA

T. Selfa and R.A. Jussaume, Jr

> What we'd like to see is people making that choice, buying organic or sustainable, or whatever, recognizing that that choice is also related to production. So it is not just for their own health, but it is seen on the landscape.

(Interview with farmland preservation advocate, Skagit County, Washington State)

> I've had customers get involved with farmland preservation ... and others interested in farm politics. But to me, I have to make them like the food. Because if they don't like the food, and they don't know how to cook it, then it doesn't hit their world.

(Interview with farmer who runs a Community Supported Agriculture (CSA) enterprise in Skagit County, Washington State)

Introduction

The environmental effects of consumption practices have recently become an area of inquiry for social scientists and policymakers in advanced industrialized countries, signalling the shift from an exclusive focus on the impacts of production on the environment (Cohen and Murphy, 2001; Princen *et al.*, 2002). This follows an extensive record of social science research on the origins of environmental values and attitudes, and whether and to what extent environmental values are reflected in environmentally friendly behaviours (Dunlap and Van Liere, 1978; Catton and Dunlap, 1980; Chalwa, 1999; DeYoung, 2000). Environmental knowledge has been increasingly identified as a foundation for environmentally friendly behaviour, because it creates awareness of environmental impacts. As with environmental orientation, knowledge about environmental problems has not been shown to necessarily correlate with environmentally friendly practices, as many factors, such as demographic and lifestyle characteristics, may intervene to shape actual consumption behaviours (Corraliza and Berenguer, 2000; Barr, 2003).

The salience of 'sustainable consumption' emerged in international environmental policy discourses following the 1992 Earth Summit, during which representatives of less economically developed countries (LEDCs) forced attention to the environmental impacts and resource intensity of northern consumption patterns (Cohen and Murphy, 2001; Princen, *et al.*, 2002; Seyfang, 2005). While attention has subsequently been given to defining sustainable consumption within the environmental policy discourses of more economically developed countries (MEDCs), much of this literature focuses on natural resource use and the capacity of the earth to absorb waste from over-consumption in the north.

Environmentalists in MEDCs advocate two predominant forms of environmentally friendly citizen-consumer behaviours - the practice of personal constraint to reduce consumption and/or adopting more sustainable or 'green' consumption practices. Thus, social science research has focused on why individuals and households reduce consumption of resources, and whether the changes in attitudes and behaviours are maintained over time. Other work has examined whether 'green' consumption practices are correlated with other types of environmentalist action and a shift toward more environmentally sustainable lifestyles (Gilg and Barr, 2005). How institutions and policies can facilitate changed behaviours, and how environmental norms and values are reflected in ordinary consumption practices, have also been areas of interest (Halkier, 2001). Environmental policy discourse, particularly in Europe, has been dominated by ecological modernization approaches to 'greening' consumption via production, through promoting the design of more energy and resource-efficient technologies and products to alter consumption patterns in a more environmentally friendly direction (Spaargaren and Van Vliet, 2000; Murphy, 2001; Carolan, 2004; Mol and Spaargaren, 2004).

In addition to creating more energy and resource efficient products for consumers, influencing environmental values related to consumption involves public education about the relationship between quality of life and consumption choices. A key challenge for policymakers is how to encourage consumers to recognize the implications of their consumption choices for natural resources and the landscape (Cohen, 2001; Murphy, 2001). In contrast to Europe and to the UK, where initiatives such as the former Countryside Agency's 'Eat the View' programme, 2000-2006 (Countryside Agency, 2002), assist consumers in making connections between food consumption patterns and environmental and landscape amenity values, similar public information campaigns linking landscape and consumption have been much less evident in North America, and certainly have not been promoted by federal government agencies on a national scale.

In recent years, prominent non-governmental organizations (NGOs) and researchers in the US have been connecting the issue of land preservation to agricultural and food production, especially at 'urban fringe' locations where land is under intense threat from development interests (American Farmland Trust, 1994; Furuseth and Lapping, 1999). Despite not being on the national policy agenda, states and regions within the US have enacted local and regional policies and new marketing initiatives to promote agricultural land preservation and agricultural revitalization, especially in urban and peri-urban regions of California, New England, the mid-Atlantic and the Pacific Northwest and other locales facing strong urban development pressures. Many of these regions have also witnessed the growth of new producer-consumer agri-food networks since the late 1990s, such as farmers markets and community supported agriculture (CSA), which provide direct market access and better prices for farmers, fresher food for consumers and the potential for community building (Lyson, 2004). Research in the New York metropolitan area has shown that while some urban consumers will pay more for 'local food' to support farmers and the regional agricultural economy, other consumers will pay more if they are made aware of the environmental and public goods provided by agriculture (Pfeffer *et al.*, 2001).

The growth of these new producer-consumer networks has been matched by an outpouring of academic scholarship assessing the reasons for their emergence, and the social and economic impacts of the networks (Lockie and Kitto, 2000; Goodman, 2002; Guthman, 2002; Raynolds, 2002; Hinrichs, 2003; Fine, 2004; Selfa and Qazi, 2005). Notably, however, the agri-food literature has been relatively silent on the environmental impacts and implications of these new producer-consumer networks. This may be due to the fact that because many of these networks operate locally, researchers assume that environmental benefits accrue from the reduction of 'food miles.'

There are a few notable examples of agri-food scholarship that have empirically assessed whether environmental values and attitudes of consumers in these new producer-consumer networks correspond with actual consumption practices. Lockie and colleagues (2002; 2004) investigated the relationship between consumption of organic food and an overall 'greening' of consumer lifestyles. Based on a project that included focus groups and a consumer survey in Australia, their research looked directly at whether interest in purchasing organic food correlates with other environmental friendly practices, pro-environment and health values, concerns about food safety and biotechnology issues, and/or particular demographic characteristics. Their research revealed consumer ambivalence about the extent to which their consumption behaviour meaningfully reflected their values about the environment, which the authors suggest is indicative of the constant redefinition of consumers' relationship to food and the contested meaning of organics (Lockie *et al.*, 2002). In addition, recent research by Seyfang (2006) examined a local organic food network in the UK as an example of sustainable consumption. Through interviews with the food network organization's founders, participant surveys, document analysis and site visits, Seyfang determined that the food network was an expression of 'ecological citizenship', because the values and practices of both founders and users of the network were resonant with notions of ecological citizenship and an environmental ethic.

Our chapter builds on this scholarship to examine the environmental attitudes and practices of consumers in Washington State, USA. In particular, we examine how consumers in four counties - two that have urban areas and two that do not - respond to several important agricultural and environmental policy statements to determine whether their environmental attitudes are reflected in how they make decisions as consumers, using data from interviews and surveys of consumers in Washington State. Because we recognize that the meanings ascribed to organic, sustainable and local food are complex and contested for consumers, our analysis goes 'beyond organic' to assess how environmental attitudes may shape alternative conceptions of what constitute 'environmentally sustainable' consumption practices. To understand the context of agri-food systems in Washington, we provide a brief contextual description and an explanation of how our data were collected.

The context of Washington State

The State of Washington lies at the extreme northwestern edge of the continental United States. It is distinguished by two major mountain ranges (the Cascades and the Olympics) that run from north to south, while the entire western border of the state is the Pacific Ocean (see Fig. 10.1). Thus, the state is characterized by a large diversity of climactic and geological conditions, ranging from temperate rain forests in the extreme western regions of the state, to an arid, central plateau region. In addition, the western part of the state is more urbanized and has a comparatively greater number of alternative food initiatives like farmers markets, urban community gardens, farmland preservation groups, CSAs and food co-operatives. In the late 1970s, the Tilth Association was formally incorporated in north-west Washington to 'support and promote biologically sound and socially equitable agriculture in the Pacific Northwest', and spawned Tilth chapters in northern California and in Oregon where it subsequently became a major West Coast organic certifying organization (Tilth Producers, 2006). In the late 1990s, the non-profit Cascade Harvest Coalition was launched to preserve and revitalize the food and farming system in western Washington through protecting farmland from urbanization, improving food marketing and promotion and facilitating the transfer of farmland from retiring to new farmers (Cascade Harvest Coalition, 2006). In 2006, the State of Washington had nearly 96 farmers markets which were members of the Washington State Farmers Market Association, of which 77 were located on the urbanized west side (Washington State Farmers Market Association, 2006). Northwest Washington is also the birthplace of Cascadian Farms, the erstwhile 'hippie' organic farm that morphed into a highly successful mainstream industrial organic company in the 1990s.

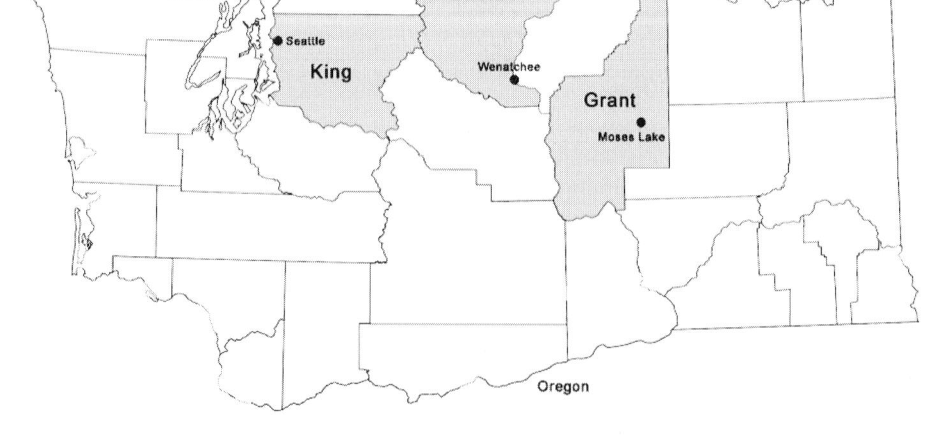

Fig.10.1: Washington State and the four study counties

By contrast, eastern Washington is predominantly rural, characterized by several large-scale agricultural commodity chains, and fewer alternative food networks have taken hold (Qazi and Selfa, 2005). The comparative lack of direct market venues, such as farmers markets, roadside stands and U-Pick operations, is notable in the heart of a highly productive agricultural landscape. Some small regional grocery stores feature seasonal 'local' produce, but there are relatively few direct marketing initiatives in eastern Washington. An interview with an alternative small farmer in eastern Washington revealed frustration with rural consumers:

> I just have a problem with rural communities... you see a lot of it, maybe we're just in too isolated of a location here... there are some sensible people but when are the rural people going to accept it? They'd much rather go to Albertson's or Safeway and not know where their food comes from.

Our analysis comparing the influence of consumer environmental attitudes on their behaviours draws data from a research project that examined various structural and local elements of the agri-food system in Washington State through in-depth interviews, focus groups and surveys of consumers and producers in the state. Following the completion of a survey of agricultural producers in the state, a telephone survey of consumers in four key counties (King, Chelan, Grant and Skagit), was implemented in the autumn of 2002. Given the challenges in obtaining a sample of reliable mailing addresses from the general population of those counties, the research team decided to utilize a random-digit dialling telephone survey for the consumer survey portion of the project. Thus, the population for the survey was all telephone households located within the four counties.[1] Ultimately, 950 respondents, with a minimum of 230 in each county, agreed to participate in the survey. The overall response rate was 23%, from a low of 21% in King County to 25% in Grant County.

Measurement procedures

In the survey we asked consumers a series of questions regarding where they purchased food and considerations that were important to them when they made purchases. Over 75% of consumers in our sample purchased food at least once a week at grocery stores. While most consumers (95%) preferred to purchase food at grocery stores, significant percentages also wanted to purchase through more direct markets: 49% highlighted farmers markets; 40% roadside stands; 20% food co-ops; 20% U-Pick operations; 16% CSAs. While there is no way to confirm that consumers actually use these options on a consistent basis, we do feel that these variables give at least an indirect indication of tendencies in food buying behaviours.

At the core of our analysis, we decided to utilize a set of measures that were based on survey question items that asked consumers whether they, when shopping, placed 'no', 'somewhat' or 'very much' importance on whether foods they purchased were produced: (a) organically; (b) in an environmentally friendly manner; and/or (c) locally. Thus, each of these variables was measured on a three point scale with '0' meaning no consideration, '1' meaning somewhat of a consideration, and '2' meaning the consumer placed very much importance on this consideration. In our sample, nearly 56% of consumers place some or very much importance on whether a crop was produced organically, while nearly 86% said that environmentally friendly production methods were important, and almost 95% said that it was important to help local farmers (see Table 10.1). It is notable that consumers were more supportive of policies to maintain and protect local family farms and farmland than were the farmers we had surveyed previously. In addition, consumers appeared to be more concerned and cautious about the environmental risks of GMOs than were farmers.

The variables that we selected and constructed for use as dependent variables in our analysis generally represent categories rather than activities that can be measured on a continuous level. In other words, the decision to consider organics, a farmer's environmental practices, or whether the source of the food is local in food buying decisions is something that either happens or not. Therefore, rather than use an Ordinary Least Squares (OLS) statistical procedure for explaining variance in the dependent variables, we employed a technique known as Ordinal Logistic Regression (McKelvey and Zavoina, 1975; Maddala, 1983). The interpretation of the coefficients generated by this technique is intuitively similar to an OLS regression technique.

Table 10.1: Distribution of attitudinal and demographic variables for consumer analyses

		N	Percentage
Less than 40 years old		337	36.12
More than 50 years old		265	28.40
Household income < $25K		199	24.09
Household income > $50K		364	44.07
4 year college degree		369	39.01
	Don't know/Not important (%)	Somewhat important (%)	Very important (%)
Importance of 'Organic'	42.16	41.27	15.57
Importance of 'Environment Friendly'	14.35	40.99	44.66
Importance of 'Help Local Farmers'	5.08	25.26	69.66
	Strongly disagree/ Disagree	Neither	Strongly agree/ Agree
Maintain family farms important	3.87	5.69	90.44
Farmers' contributions appreciated	22.63	5.98	71.40
Risks of GMOs not well understood	9.05	3.39	85.56
Farmers should be paid to protect	11.14	6.80	81.97
Govt should restrict non-agric devt	12.92	4.94	82.13

For use as independent variables in our analysis, we chose several demographic and attitudinal variables. For our demographic variables, we settled on age, income and education. Such variables have often been used to explain behaviours, and we wanted to contrast the influence of demographic variables with attitudinal variables. For age, we differentiated between consumers who were less than 40 years old (36%) and 50 years old or more (28%). For income, we distinguished between those whose total incomes were less than US$25,000 (24%) and more than US$50,000 (44%) per year. Finally, for education, we created a dummy variable that distinguished between the 39% of consumers who had completed a 4 year course of post-secondary formal education and those who had not.

One of our main objectives in this chapter is to examine whether the attitudes of consumers towards various environmental and agricultural policy issues helped to explain the factors they considered when purchasing food. Therefore, we also asked consumers a series of agricultural and environmental policy questions (see Table 10.1). The environmental and agricultural policy variables which we chose to use in our analysis asked respondents the degree to which they agreed or disagreed with the following statements:

- Maintaining family operated farms is important to the future of my county.
- Farmers' contributions to land stewardship are appreciated by the general public.
- The environmental risks of genetically modified crops are not well understood.
- Farmers should be paid for their participation in wildlife programmes including those to protect habitat.
- Local government should restrict non-agricultural development in important agricultural areas.

In all cases, we found that a fairly high percentage of consumer respondents strongly agreed with these statements (see Table 10.1 for percentages). Therefore, we decided to create a series of dummy variables that differentiated between respondents who did or did not strongly agree with the statement in question.

Results and discussion

We utilized demographic indicators and responses to agricultural and environmental policy questions in three models to analyse consumer purchasing patterns. In particular, we examined how well these demographic and attitudinal factors predict which consumers claim to be interested in buying organic foods, environmentally-sustainably produced foods, and foods that support local farmers. In Table 10.2, we can see that none of the demographic variables, nor any of the policy variables, help explain whether consumers consider if a product is organic when purchasing food, with the exception of support for government restrictions on non-agricultural development, which was statistically significant at a level of $P<.001$.

What this finding reveals is consumers in our sample who expressed very strong concerns about the environmental risks of GMOs were *not* more likely than consumers who did not share these concerns to consider organics as an important factor in their food purchasing decisions, nor did support for the other environmentally oriented policy statements correlate with an interest in buying organic food. In addition, we believe it is important to highlight that neither consumer affluence nor higher education levels, factors which have been linked to interest in organic purchases in other research (Torjusen *et al.*, 2001; Guthman, 2003; Goodman, 2004), were significant in our analysis. Our finding of no causal link between demographic or attitudinal factors, other than the belief that government should restrict non-agricultural development in important agricultural regions, and organic purchasing supports other research that suggests that because consumers must manage competing discourses of environmental and health benefits, price, convenience and availability when making food purchases, the value or meaning of organic foods may not be sufficiently clear and compelling (Lockie *et al.*, 2002; 2004). Data from our focus group discussion with CSA members supported this interpretation that purchasing decisions for consumers are not clear cut because there are competing discourses of quality, price and environmental friendliness to assess. When asked why they joined a CSA and how important it was to have their CSA based exclusively around organic products, CSA members expressed varied sentiments which expose some of the tensions they face.

Table 10.2: Ordered Logit Model estimates for whether organics influence consumers' food buying decisions

Variable	Full Model			Restricted Model		
	Coef. (Std. Error)	z	P>[z]	Coef. (Std. Error)	z	P>[z]
Less than 40 yrs old	.992 (.1778)	0.56	0.577	---	---	---
More than 50 yrs old	-.1740 (.1973)	-0.88	0.378	---	---	---
Household income < $25K	.2137 (.2150)	0.99	0.320	---	---	---
Household income > $50K	.0092 (.1726)	0.05	0.957	---	---	---
4 year college degree	.0928 (.1591)	0.58	0.560	---	---	---
Maintain family farms important	-.0033 (.1765)	-0.02	0.985	---	---	---
Farmers' contributions appreciated	-.0018 (.1624)	-0.01	0.991	---	---	---
Risks of GMOs not well understood	.1315 (.1584)	0.83	0.407	---	---	---
Farmers should be paid to protect	.1149 (.1622)	0.71	0.479	---	---	---
Govt should restrict non-agric devt	.7359 (.1643)	4.48	0.000	.7054 (.1294)	5.45	0.000
N	650			880		
Wald Chi-Square	31.36			30.19		
Prob > Chi-Square	0.0005			0.00		
Degrees of Freedom	10			1		

One of the CSA members, an entomologist, explained that he took issue with the safety of some of the sprays that are allowable under the organic label because, while they may be safe for human consumption, they harm beneficial insect populations:

> On the organic question, I would like to support organics also but I am probably more interested in sustainable because there are probably some practices that have been done conventionally that can be incorporated into a system that can be quite viable.... Philosophically I just don't align myself very closely with organic ... I like the idea of small, I like the idea that it promotes a better understanding of the system and how it works.

Another member who had only recently become aware of the existence of community supported agriculture joined the local CSA mainly because of the convenience; organic was a secondary consideration for her:

> Someone brings these to me and I don't have to go the farmers' markets. That's why I joined [the CSA] ... I don't always buy organic, but if it's available, sure.

Another member had wanted to join a CSA for many years but one had not been available locally. For her, organic agriculture was a value she wanted to support:

> I was really excited to get some of their produce, because it's organic and carefully raised. So I am always looking for ways that I can do that, partly because I think that everything that I buy is a vote. To me, one of the benefits is not just that the food is really good, but I feel part of a larger vision that is really important to me of the picture of the Wenatchee Valley becoming a sustainable place, a green valley where the soil is restored - it's no longer that we have to take away the soil and bring in new stuff to plant trees!

Yet another respondent mentioned that while she had never been in a CSA previously, her family had always been avid organic home gardeners before it was labelled as such. She said:

> The organic focus is not crucial to me ... I do like the fact that there is no biotechnology. To me it's more a question of an increase in consumption of vegetables. I am coming at it from a more nutrition point of view that any intake of whole foods, instead of a vitamin, is cancer preventative and all sorts of things like that.

Among other concerns, consumers in the focus groups expressed desires to balance nutrition and health attributes with interest in promoting understanding of sustainable systems and supporting smaller scale farming. It is notable that consumers in the CSA differentiate between organic and sustainable food, suggesting that their conception of sustainable may be broader than the organic label. For others, buying organic is a means of ensuring that other important values are met, such as not buying food produced with biotechnology.

Table 10.3: Ordered Logit Model estimates for whether environmentally friendly practices influence consumers' food buying decisions

Variable	Full Model			Restricted Model		
	Coef. (Std. Error)	z	P>[z]	Coef. (Std. Error)	z	P>[z]
Less than 40 yrs old	.3019 (.1851)	1.63	0.103	---	---	---
More than 50 yrs old	-.0368 (.2101)	-0.18	0.861	---	---	---
Household income < $25K	.4017 (.2345)	1.71	0.087	.4544 (.2192)	2.07	0.038
Household income > $50K	-.3291 (.1819)	-1.81	0.070	-.3217 (.1749)	-1.84	0.066
4 year college degree	-.0826 (.1667)	-0.50	0.620	---	---	---
Maintain family farms important	.4093 (.1808)	2.26	0.024	.3970 (.1706)	2.33	0.020
Farmers' contributions appreciated	.2713 (.1730)	1.57	0.117	---	---	---
Risks of GMOs not well understood	.5834 (.1655)	3.53	0.000	.6005 (.1602)	3.75	0.000
Farmers should be paid to protect	.3990 (.1702)	2.34	0.019	.4482 (.1618)	2.77	0.006
Govt should restrict non-agric devt	.8430 (.1707)	4.94	0.000	.7897 (.1631)	4.84	0.000
N	646			681		
Wald Chi-Square	107.79			103.32		
Prob > Chi-Square	0.00			0.00		
Degrees of Freedom	10			6		

By contrast, in the second model, a number of the independent demographic and policy variables we selected *do* explain why consumers have an interest in purchasing food that is produced in an 'environmentally friendly manner' (see Table 10.3), a description which one might argue is more easily interpretable by consumers, although more difficult for farmers to define and implement. Income was strongly correlated with an interest in environmentally friendly food. Interestingly, respondents from households with total incomes of less than US$25,000 a year were slightly *more* likely to consider environmental friendliness in their food purchasing decisions, while those from households with incomes of more than US$50,000 a year were slightly *less* likely to (only the less than US$25K variable was significant at $P<.05$ in the restricted model). This certainly appears to contradict the assumption that it is wealthier consumers who are more interested in supporting environmentally friendly agriculture through buying 'yuppie chow' (Guthman, 2003). In addition, for consumers who responded to our survey, positive responses to nearly all the pro-farming and environmental policy variables we included in our model were significantly associated (often at a level of $P<.001$) with a greater likelihood of considering environmentally friendliness when buying food. There was a strong correlation between believing the government should protect important agricultural land from development and an interest in purchasing environmentally friendly food. Clearly, choosing to buy food because it is produced in an 'environmentally friendly manner' has more resonance and meaning to consumers who are supportive of policies to maintain local farms and concerned about environmental issues than does purchasing organic food. We might also suggest that 'environmentally friendly production' encompasses a much greater spectrum of practices than does the organic label, and therefore leaves more room for consumers to make judgments about how products might connect their values about the environment with their purchasing decisions.

Table 10.4: Ordered Logit Model estimates for whether wanting to help local farmers influences consumers' food buying decisions

Variable	Full Model			Restricted Model		
	Coef. (Std. Error)	Z	P>[z]	Coef. (Std. Error)	z	P>[z]
Less than 40 yrs old	-.5808 (.2228)	-2.61	0.009	-.4560 (.1798)	-2.54	0.011
More than 50 yrs old	.1986 (.2773)	0.72	0.474	---	---	---
Household income < $25K	.1570 (.2829)	0.55	0.579	---	---	---
Household income > $50K	-.2327 (.2222)	-1.05	0.295	---	---	---
4 year college degree	-.4754 (.2007)	-2.37	0.018	-.7124 (.1791)	-3.98	0.000
Maintain family farms important	1.5355 (.2042)	7.52	0.000	1.6484 (.1891)	8.72	0.000
Farmers' contributions appreciated	.3822 (.2170)	1.76	0.078	.2012 (.1996)	1.01	0.313
Risks of GMOs not well understood	.1640 (.1987)	0.83	0.409	---	---	---
Farmers should be paid to protect	.4532 (.2042)	2.22	0.026	.3628 (.1871)	1.94	0.053
Govt should restrict non-agric devt	.4159 (.2049)	2.03	0.42	.6139 (.1860)	3.30	0.001
N	652			760		
Wald Chi-Square	144.72			169.88		
Prob > Chi-Square	0.00			0.00		
Degrees of Freedom	10			6		

The final model (see Table 10.4) looked at the importance of wanting to support local farmers through consumer food purchase decisions. In a sense, this statement was testing whether consumers had awareness of and interest in supporting a more local food system which has been positively associated in the agri-food literature with greater environmental and economic sustainability, but also conversely with a parochial defence of local farms without any commitment to environmental sustainability or social inclusion (Hinrichs, 2003; Winter, 2003; Qazi and Selfa, 2005). In this model we find that, of the demographic variables, age and education are significant predictors of those consumers who value local farmers: older consumers over the age of 40 and those who had not completed a 4 year college degree programme were more interested in supporting local farmers through their purchases. Concerns about the environmental risks of GMOs were not linked to preference for supporting local farms through food purchases, whereas favouring paying farmers to protect wildlife habitat was weakly linked (at a level of $P<.05$ in the full model) to preferences for helping local farmers. Not surprisingly, being strongly in favour of policies to maintain local family farms and to restrict non-agricultural development were highly significant predictors of food purchasing decisions that are directed towards supporting local farmers, particularly in the restricted model. In this case, 'local foods' discourse may be linked to a politics of 'defensive localism' rather than a turn toward quality production centred around organic and/or ecologically sustainable production practices (Winter, 2003).

Conclusions

The purpose of this chapter was to examine whether demographic characteristics and consumers' attitudes toward agriculture, farming and the environment could be linked to particular consumption practices, using interview and survey data from the state of Washington. While there has been a great amount of social science scholarship devoted to understanding the relationship between environmental attitudes and practices, there has been little attention to whether consumers' food purchasing behaviour is consistent with their attitudes toward the environment and farming. Previous related research has only focused on the relationship between consumer environmental attitudes and their interest in buying *organic* food, but our analysis of environmental attitudes and behaviour considered a broader definition of sustainability to include consumer interest in purchasing food produced in an environmentally friendly manner and foods produced locally.

This analysis has shown that for consumers in four counties in Washington State, there was a strong correlation between wanting to preserve farmland and buying organics, but attitudes toward other environmental policies and/or demographic variables were not linked with a preference for organic products. However, buying 'environmentally friendly produced' foods had more resonance and meaning for consumers with pro-environment and agriculture values. The strongest predictor for buying 'environmentally friendly produced' food was support for policies to preserve farmland. Income levels were also strongly linked to making environmentally friendly purchases, but in the reverse direction to that suggested in the literature: lower income households were *more* likely and higher income households were *less* likely to consider whether food was produced in an environmentally friendly manner when making purchasing decisions. Age was not a factor for either the decision to buy organic or environmentally friendly foods, but older consumers were more interested in buying 'local' food as a way to support local farmers. But the strongest predictor for consumers who considered helping local farmers when making purchases was their support for policies to maintain local family farms.

It is clear that for consumers, environmental attitudes are not uniformly associated with purchasing or eating local, organic and/or environmentally sustainably produced food. However, the one variable that was significant in all three models was respondent interest in government restrictions on non-agricultural development in important agricultural areas. This indicates that consumers place a very strong value on the societal benefits provided by farmers' agri-environmental stewardship, suggesting that consumer purchasing decisions are not limited to, although certainly incorporate, the maximization of their individual utility.

The policy implications of this research are also important. We demonstrate that consumers appear to support government policies restricting the expansion of urban development and protecting farmland. Up to now it has been primarily citizens' groups and non-governmental organizations in western Washington, which is facing pressures of rapid urbanization and suburbanization, which have recognized the threat of farmland loss and have actively promoted public education about the importance of farm landscapes and food production. These civil society organizations have been working on improving farmer-consumer marketing links and on forging economically viable inter-generational transfers of farmland. Our research indicates that consumer-citizens in Washington State favour a more active government role in preserving farmland. This suggests that while NGOs have taken the lead in fostering farmland preservation through market approaches, citizen-consumers believe there is also a role for the state in creating sustainable foodscapes.

Acknowledgements

We would like to acknowledge the financial support of the United States Department of Agriculture, the W.K. Kellogg Foundation and the Washington Farming and the Environment Project for the research on which this chapter is based. We would also like to thank Lucy Jarosz, Department of Geography, University of Washington, who collected the interview data in Skagit County and Joan Qazi, Wenatchee Valley College, who collected the focus group data in Chelan County.

Notes

1. Genesys, Inc. prepared the sampling frame for the consumer survey. Telephone numbers were generated randomly using a computer after determining all the working exchanges and working blocks within the counties. A random sample of 5200 telephone numbers, with 1300 in each target county, was selected. Of these 5200 telephone numbers, 1043 numbers were determined to be business and/or non-working numbers and were purged. This made the corrected sample 4157 numbers. Telephone calling began on October 15th 2002 and continued through December 19th 2002. Interviewers asked to speak with the person living in this household, 18 years or older, who was most involved with food buying for the household. A minimum of 12 call attempts was made to each number.

References

American Farmland Trust (1994) *Farming on the Edge: A New Look at the Importance and Vulnerability of Agriculture near American Cities*. American Farmland Trust, Washington, DC, USA.

Barr, S. (2003) Strategies for sustainability: citizens and responsible environmental behaviour. *Area* 35(3) 227-240.

Carolan, M. (2004) Ecological modernization theory: what about consumption? *Society and Natural Resources* 17, 247-260.

Cascade Harvest Coalition (2006) Cascade Harvest Coalition, Seattle, Washington, USA. [Accessed June 2006] Available from http://www.cascadeharvest.org

Catton, W. and Dunlap, R. (1980). A new ecological paradigm for post-exuberant sociology. *American Behavioral Scientist* 24(1) 15-47.

Chalwa, L. (1999) Life paths into environmental action. *Journal of Environmental Education* 31(1) 15-27.

Cohen, M.J. (2001) Emergent environmental policy discourse on sustainable consumption. In: Cohen, M.J. and Murphy, J. (eds) *Exploring Sustainable Consumption: Environmental Policy and the Social Sciences*. Pergamon, Oxford, UK, 21-37.

Cohen, M.J. and Murphy, J. (eds) (2001) *Exploring Sustainable Consumption: Environmental Policy and the Social Sciences*. Pergamon, Oxford, UK.

Corraliza, J. and Berenguer, J. (2000) Environmental values, beliefs and actions: a situational approach. *Environment and Behavior* 32(6) 832-848.

Countryside Agency (UK) (2002) *Eat the View: Promoting Sustainable Local Products*. Countryside Agency Publishing, Wetherby, UK.

DeYoung, R. (2000) New ways to promote proenvironmental behavior: expanding and evaluating motives for environmentally responsible behavior. *Journal of Social Issues* 56(3) 509-526.

Dunlap, R. and Van Liere, K. (1978) The new environmental paradigm. *Journal of Environmental Education* 9(4) 10-19.

Fine, B. (2004) Debating production-consumption linkages in food studies. *Sociologia Ruralis* 44(3) 332-342.

Furuseth, O. and Lapping M. (eds) (1999) *Contested Countryside: The Rural Urban Fringe in North America*. Ashgate, Hants, UK.

Gilg, A.W., Barr, S. (2005) Green consumption or sustainable lifestyles? Identifying the sustainable consumer. *Futures* 37(6) 481-504.

Goodman, D. (2002) Rethinking food production-consumption: integrative perspectives. *Sociologia Ruralis* 42(4) 271-277.

Goodman, D. (2004) Rural Europe redux? Reflections on alternative agro-food networks and paradigm change. *Sociologia Ruralis* 44(1) 3-16.

Guthman, J. (2002) Commodified meanings, meaningful commodities: re-thinking production-consumption links through the organic system of provision. *Sociologia Ruralis* 42(4) 295-311.

Guthman, J. (2003) Fast food/organic food: reflexive tastes and the making of 'yuppie chow'. *Social and Cultural Geography* 4(1) 45-58.

Halkier, B. (2001) Routinisation or reflexivity? Consumers and normative claims for environmental consideration. In: Gronow, J. and Warde, A. (eds) *Ordinary Consumption*. Routledge, London, UK, 25-44.

Hinrichs, C. (2003) The practice and politics of food system localization. *Journal of Rural Studies* 16(3) 33-45.

Lockie, S., Lyons, K., Lawrence, G. and Grice, J. (2004) Choosing organics: a path analysis of factors underlying the selection of organic food among Australian consumers. *Appetite* 43, 135-146.

Lockie, S., Lyons, K., Lawrence, G. and Mummery, K. (2002). Eating 'green': motivations behind organic food consumption in Australia. *Sociologia Ruralis* 42(1) 23-40.

Lockie, S. and Kitto, S. (2000) Beyond the farm gate: production-consumption networks and agri-food research. *Sociologia Ruralis* 40(1) 3-19.

Lyson, T. (2004) *Civic Agriculture: Reconnecting Farm, Food and Community*. Tufts University Press, Medford, Massachusetts, USA.

Maddala, G.S. (1983) *Limited-Dependent and Qualitative Variables in Econometrics*. Cambridge University Press, Cambridge, UK.

McKelvey, R.D. and Zavoina, W. (1975) A statistical model for the analysis of ordinal level dependent variables. *Journal of Mathematical Sociology* 4, 103-120.

Mol, A.P.J. and Spaargaren, G. (2004) Ecological modernization and consumption: a reply. *Society and Natural Resources* 17, 261-265.

Murphy, J. (2001) From production to consumption: environmental policy in the European Union. In: Cohen, M.J. and Murphy, J. (eds) *Exploring Sustainable Consumption: Environmental Policy and the Social Sciences*. Pergamon, Oxford, UK, 39-58.

Pfeffer, M.J., Stycos, J.M., Glenna, L. and Altobelli, J. (2001) Forging new connections between agriculture and the city. In: Solbrig, O., Paarlberg, R. and DiCastri, D. (eds) *Globalization and the Rural Environment*. Harvard University Press. Cambridge, Massachusetts, USA, 419-446.

Princen, T., Maniates, M. and Conca, K. (eds) (2002) *Confronting Consumption*. MIT Press, London, UK.

Qazi, J. and Selfa, T. (2005) The politics of building alternative agro-food networks in the belly of agro-industry. *Food Culture and Society* 8(1) 45-72.

Raynolds, L.T. (2002) Consumer/producer links in Fair Trade coffee networks. *Sociologia Ruralis* 42(4) 404-424.

Selfa, T. and Qazi, J. (2005) Place, taste or face-to-face? Understanding producer-consumer networks in "local" food systems in Washington State. *Agriculture and Human Values* 22, 451-464.

Seyfang, G. (2005) Shopping for sustainability: can sustainable consumption promote ecological citizenship? *Environmental Politics* 14(2) 290-306.

Seyfang, G. (2006) Ecological citizenship and sustainable consumption: examining local organic food networks. *Journal of Rural Studies* 22, 383-395.

Spaargaren, G. and Van Vliet, B. (2000) Lifestyles, consumption and the environment: The ecological modernization of domestic consumption. *Environmental Politics* 9, 50-77.

Tilth Producers (2006) Tilth Producers of Washington, Seattle, USA. [Accessed June 2006] Available from http://www.tilthproducers.org

Torjusen, H., Lieblein, G., Wandel, M. and Francis, C.A. (2001) Food system orientation and quality perception among consumers and producers of organic food in Hedmark County, Norway. *Food Quality and Preference* 12, 207-216.

Washington State Farmers Market Association (2006) Washington State Farmers Market Association, Suquamish, Washington, USA. [Accessed 2006] Available from http://www.wafarmers markets.com

Winter, M. (2003) Embeddedness, the new food economic and defensive localism. *Journal of Rural Studies* 19, 23-32.

Part III

Ethical Production and Protection

Chapter 11

A notional ethical contract with farm animals in a sustainable global food system

B. Mepham

Introduction

The aim of this chapter is to argue for a new ethical theory to inform and justify a continued use of animals in the global food system. It is however important to appreciate that, if put into practice, the theory to be outlined would entail a major change in our treatment of most farm animals. In no sense should the reasoning employed be construed as a defence of the currently dominant systems of animal production.

The late Bernard Williams defined a theory of ethics as 'a philosophical structure, which together with some degree of empirical fact, will yield a decision procedure for moral reasoning' (Williams, 1993: 6). It is an important point - which deserves to be emphasized - that changes in our objective understanding (or in Williams' terms, a new appreciation of 'empirical facts') often demand reassessment of traditional norms of behaviour. In the present context, recent progress in genomics has provided a wealth of evidence in support of Darwin's theory of evolution. This establishes, beyond any reasonable doubt, the genetic continuity of human and non-human species. As graphically expressed by Richard Dawkins:

> But for the accidental extermination of the intermediates linking us to, for example, chimpanzees, we should be united to them by an interlocking chain of interbreeding: a daisy chain of the 'I've danced with a man, who's danced with a girl, who's danced with the Prince of Wales' variety (Dawkins, 1999: xv-xvi).

And if further proof were needed, we have only to consider the programmes which aim to use pig organs, with but minor genetic modification in the form of xenografts, as replacements for defective human organs.

So, the philosophical question we are now faced with, which was barely raised before Darwin, and can now hardly be avoided, is 'What is an animal?'(Ingold, 1994). It is not possible to explore that important question, which I have recently addressed elsewhere (Mepham, 2006) in any detail in this chapter. But such are the similarities between humans and many non-humans that it needs only a little reflection to conclude that it might be better to turn the question on its head and try to define what a 'person' is. The usual line of argument emphasizes characteristics thought to be unique to humans, such as the abilities to use language, employ reasoning and deploy tools. But the difficulty with such criteria is that some people (such as infants, some very old and some disabled people) *lack* these abilities, whereas some animals appear to *demonstrate* them.

Even if major differences between animals and humans are conceded, some philosophers have argued that in *ethically relevant* respects animals do not differ substantially from humans, or at least do not differ in ways that would justify killing animals for food. Two main lines of reasoning have received much attention - the animal rights argument, prominently articulated by Tom Regan (Regan, 1985) and a form of utilitarian argument, which reckons the nutritional benefits and pleasure to humans as greatly outweighed by animal suffering, as argued by Peter Singer. For Singer: 'Pain is pain, whatever the species of being experiencing it' (Singer, 1990: 6). That these philosophical arguments are gaining ground would seem to be borne out by several changes evident in society (at least in UK and EU society). For example: (i) legal measures to protect animal welfare, both on farms and in laboratories, have become more stringent in recent years; (ii) hunting with dogs became illegal in the UK in 2005; and (iii) there appears to be a steady increase in the numbers of people claiming to be vegetarians.

So a question to which all humane, rational people seem now obliged to give serious attention is whether, given our current understanding of the nature of non-human animals, it is justifiable, in ethical terms, to use them in our agricultural systems, and in particular eat their flesh when this is procured by slaughtering them.

Utilitarian justifications for using animals as sources of food

Utilitarianism, first formulated by Bentham and Mill (Mill, 1910) in 18th and 19th century England, is a consequentialist theory which identifies good actions as those which maximize benefits and minimize costs: in Bentham's terms to 'produce the greatest good for the greatest number' (Bentham, 1948). This 'cost-benefit' theory forms the major justification for non-human animal use, since it would be difficult to argue that we have a *duty* to use animals, except perhaps in the probably rare cases where the *only* option entailed human suffering. It is therefore useful to begin with a consideration of the benefits which are claimed to justify animal use.

Nutritious, appetising food

Animal products, such as meat, milk, eggs and fish, are accessible sources of certain dietary nutrients which are less readily available (if at all) from vegetable sources. These include: certain essential amino acids and long-chain fatty acids, vitamins A, B12 and D, iodine, iron, selenium and zinc (Food Ethics Council, 2001). Some non-meat animal products are also rich sources of essential nutrients, e.g. milk provides bioavailable calcium and phosphorus.

Evidence that animal products satisfy the human appetite is provided by their increased consumption when they become more affordable, as illustrated by the direct relationship in less economically developed countries (LEDCs) between rising incomes and the consumption of animal products. For example, between 1967 and 1997 meat consumption increased by 118% in LEDCs as a whole, and by 343% in East Asia (Hall *et al.*, 2004).

Sustainable farming systems

Traditionally, and currently in systems such as organic farming in more economically developed countries (MEDCs), and in many agricultural systems in LEDCs, animals play key roles in farming. Ruminants, such as cattle and sheep, can bring into productivity land which is otherwise too poor, erodible or difficult to cultivate (e.g. on steep hillsides). Moreover, by converting fodder such as grass, inedible by humans, into nutritious foods such as meat and milk, they can sustain human populations on land which could not support them on plant foods alone. Grazing plays an important role in recycling plant material, increasing plant biomass and diversifying plant communities.

Animal manure also often serves important functions. For example, it increases soil fertility and is colonized by large numbers of invertebrates, which serve as a source of food for birds and mammals. The result is that local ecological systems can support rich wildlife populations in addition to the livestock and human populations. Often, the resulting pastoral landscape, e.g. in the UK, is a much valued element of the national heritage, which apart from its aesthetic qualities provides an important recreational amenity and is a significant source of tourist revenue.

Economics and employment

It is obvious from the above considerations that animals play a significant role in national economies, generating incomes and employment for farmers, food processors and associated industries such as livestock markets and veterinary drug manufacturing.

Animals also perform many non-food roles, especially in LEDCs. These include: agricultural traction power, general transportation, and provision of fibre (e.g. in the forms of wool, leather and bone). Moreover, animal dung is frequently used as a fuel for cooking and heating. Indeed, it is in LEDCs that animals play their most vital roles. According to the International Livestock Research Institute (2004: 9), 'livestock in developing countries contribute up to 80% of agricultural GDP, and some 600 million people rely to a significant degree on livestock for their livelihoods'. Moreover, the role of animal products is predicted to become even more important in future, because consumption of meat is expected to grow annually by 3.5% and milk by 3.8%, at least up to the year 2020. By then, LEDCs will produce 60% of the world's meat and 52% of the world's milk (International Livestock Research Institute, 2004).

There is a common tendency for those in MEDCs to overlook the nutritional and economic role of fish. Globally, the almost 35 million people who are engaged in fishing and fish farming as a full-time or part-time occupation are concentrated in Asia, where about 85% live. In many Asian countries, artisanal fishing families are among the most socially, economically and politically disadvantaged segments of the population, with a status comparable to landless labourers. Even so, small-scale fishing families often fare better than others, even if their average income levels are below official poverty levels (Food and Agriculture Organization (FAO), 2005: 8).

The above advantages all affect people, but some benefits of keeping farm animals might be considered to accrue to the animals themselves. These result from what has been characterized as the *co-evolution* of animals and humans. Challenging what he sees as the popular myth according to which primitive people dominated certain wild animals and forced them into domestication, Stephen Budiansky argues that rather than being an act of human exploitation, domestication was a natural process. In short, 'in an evolutionary sense' modern farm animals can be said to have 'chosen us as much as we chose them'; and the process of domestication has not only entailed profound changes in the domesticated animals but also significant changes in humans (Budiansky, 1992: 24).

Thus, to a large degree modern farm animals have become dependent on humans, who provide them with food, shelter, and protection from both climatic extremes and predators. In the course of domestication, the animals have undergone a process of juvenilization, so that in the adult state they retain many of the juvenile anatomical and behavioural features of their wild ancestors. The condition, called *neoteny*, makes them more amenable to human interactions, so that, for example, cattle yield their milk to a mechanical stimulus as well as to the suckling calf.

But humans have also become adapted to the animals with whom they have co-evolved. A striking example is the mutation in northern Europeans (and their descendents, for example, in North America and Australia) which permits them to continue during adulthood to digest the milk sugar lactose, a capability lost in most humans at the time of weaning. The persistence of the intestinal enzyme *lactase* in Europeans allows them to benefit from the rich supply of calcium in milk (important for bone and teeth formation), which may compensate for the poor absorption of dietary calcium due to low levels of vitamin D that is elsewhere in the world produced amply in skin under the influence of sunlight.

Cultural significance

Such examples of the symbiotic relationship between farm animals and humans have led to the suggestion that human–animal interactions are a vital ingredient of human culture - which has shaped our worldviews. Thus, it has been claimed that: 'We are enough like animals to be kept humble; we are different enough from animals to be aware of our unique responsibility as husbandmen of the natural world' (Hodges, 1999: 1-6). Undeniably, farm animals have been part of human culture, and for many they contribute to a commonly perceived 'natural order.'

According to this account, the mutual benefits of animal agriculture might in theory result in an outcome in which the lot of *both* humans and animals is improved by the practice of domestication.

Attitudinal and legal justifications for animal use in agriculture

In addition to the arguments advanced above, there has been a long standing tradition that farm animals are justifiably regarded as 'agricultural products' - an attitude that can perhaps be traced back to the theory of property rights enunciated by the 17th century philosopher, John Locke. While it was then commonly believed that all people were created and belonged to God, Locke argued that individuals could acquire property by joining their labour with a commonly owned object in nature. Thus, by felling a tree and constructing from the wood a piece of furniture, one could claim exclusive ownership of the furniture, which represented the fruits of one's labour. By the same token, if animals were hunted and captured from the state of nature they became the hunter's personal property, and could legitimately be sold to others for food or other purposes. For Locke: 'The labour that was mine, removing them out of that common state they were in, hath fixed my property in them' (Locke, 1924: 130).

The result was that farm animals came to be seen as valuable in a purely instrumental sense, downgrading any intrinsic value that they might also have as sentient beings (Mepham, 2006). Historical evidence reveals that the domestication and ownership of animals were closely related to the very ideas of property and money - a point strikingly made by the fact that the word 'cattle' comes from the same route as the word 'capital,' and the two are synonymous in many European languages (Francione, 2000).

The idea that farm animals are essentially 'property,' and primarily of instrumental value, led to the emergence by the mid-20th century of industrialized systems of animal production that critics described as 'factory farming' (e.g. Harrison, 1964). It has been argued that in some respects animals kept in such intensive systems benefit by comparison with others reared more traditionally, for example, in being protected from climatic extremes and attacks by predators, and in receiving diets finely matched to nutritional requirements.

Utilitarian constraints on animal use in agriculture

However, if we take the mental experiences of animals at all seriously, our current understanding of the extent to which they are likely to experience pleasure, pain and distress must lead us to question whether, especially in many intensive animal production systems, any benefits for animals are not usually outweighed by the costs to their welfare.

Negative impacts on farm animal welfare may be experienced at various points in the 'animal products food chain' which encompasses breeding, feeding, housing, transport and slaughter (Mepham, 2005: 170). For example, in the pursuit of economic efficiency, animals may be:

- Bred to grow too rapidly (as happens with some broiler chickens whose legs are unable to bear their weight).
- Fed diets which may cause digestive problems and joint disease (as is the case with some dairy cattle).
- Housed in ways which prevent normal behavioural patterns (as is notably the case for laying poultry kept in battery cages).
- Subjected to mutilations to facilitate animal management (such as castration, ear notching, debeaking and tail docking).
- Transported long distances in stressful circumstances to markets or abattoirs; and slaughtered in conditions which are not always free of mishap and consequent suffering (Food Ethics Council, 2001).

Moreover, there are other significant drawbacks, particularly of intensive animal production systems, such as BSE, foot and mouth disease and avian influenza (Mepham, 2004) and the problem of disposing of huge volumes of animal excreta, which often far exceed the local ecosystem's absorptive capacity.

Recognizing both these facts and the increasing evidence for their close genetic, physiological and psychological similarity with humans, can we continue to use (some would say abuse) animals in these ways? Ultimately, according to this utilitarian reasoning, it is of course a matter of human judgement. Decisions need to be made on the scope of the cost-benefit analysis that underpins utilitarian reasoning, and on the weight which is attached to the pleasures, pains and preferences of farm animals when stacked against the same criteria for humans. One fact that cannot be ignored is that many people who adopt veganism appear to live long and healthy lives.

Economic and environmental constraints on modern animal production systems

Of course, ethical issues concerning farm animals extend far beyond considerations of how the individual animals are reared, raised, transported and slaughtered. They relate, for example, to (i) impacts on the environment of the systems employed for feeding, housing and disposal of excreta and (ii) the socio-economic, political and legal arrangements which increasingly govern global trade in food and agriculture, and will become acute if genetically modified (GM) and cloned farm animals are introduced. For present purposes, a few key issues only are highlighted.

Environmental impacts of the drive for increased agricultural productivity are reflected graphically in biodiversity statistics. For example, a report of the UK's overseas development ministry in 2001 stated that:

> 70% of the world's rural poor rely on livestock, many of which are adapted to local conditions and diseases. [But] with a third of breeds threatened by extinction, there is a major risk to the financial security of the poorest families (Department for International Development, 2001: 4).

According to the FAO, at least 1000 breeds have been lost in the last 100 years, and the rate of loss is accelerating, with 300 breeds disappearing in the last 15 years. It is estimated that currently 2000 breeds are at risk (FAO, 2000). Intensive animal farming is a major source of water pollution, with cattle and other livestock accounting for twice the amount of pollutants that come from all other industrial sources. Cattle and sheep are also among the principal sources of methane, a greenhouse gas which ranks second to carbon dioxide in terms of its contribution to global warming (Gold, 1999).

Although attempts to introduce the techniques of genetic manipulation in farm animals have lagged behind developments in crops, the commercial pressure for such innovations is evident. Thus, John Hodges, the former head of animal breeding and genetic resources at the FAO, claimed that multiple patent applications by leading biotechnology companies in many countries in 2005 would, if successful, give the companies rights over many animal management systems which have been designed by publicly-supported scientists over many years. Hodges claims that the resulting creation of monopolistic trade arrangements within the food chain, protected by intellectual property rights, poses threats to species diversity, public acceptability (since there are widespread reservations about GM applied to animals) and public health (because a prudent application of the Precautionary Principle would pay due regard to risks of genomic interactions between human and non-human species) (Hodges, 2005).

A notional ethical contract with animals

Those who adopt an animal rights position, such as the philosopher Tom Regan, find no justification for taking the lives of other sentient beings merely to satisfy gustatory appetite.

For him, farm animals 'have a life of their own, of importance to them apart from their utility to us. They have a biography not just a biology; they are somebody not something' (Regan, 1990: 8). But the difficulty with absolutist positions is that they can lead to some highly problematical consequences.

The consequences of a global adoption of veganism would depend to a large degree on how it was implemented. In a 'fantasy scenario' where the global population of over 6 billion became vegans overnight, and all caged and penned animals were 'set free,' the negative effects on both animal and human welfare would be devastating. Without employment, farmers, fishers, butchers, livestock hauliers and their dependents would face financial ruin, while most animals would probably die hungry and fearful in inhospitable environments - and doubtless cause a public health crisis. But, probably, no one is realistically proposing that we should all become vegans overnight. The more challenging question concerns whether veganism could be phased in to result in a sustainable global food system, capable of supporting 8-10 billion people into the foreseeable future.

It would seem, however, that, for the following reasons, such an objective is at best naïve. Returning to Bernard Williams' (1993) definition of an ethical theory, it is worth rehearsing some 'empirical facts'. With high degrees of probability, farm animals:

- Have become, as a result of selective breeding, markedly dependent on human care (e.g. for food).
- In certain respects lead safer, healthier lives than their wild relatives (because they are kept in protected environments and receive veterinary treatment).
- Would often be unable to survive in the 'wild', even if such conditions existed to an adequate extent.
- Are, in a significant number of cases, an indispensable part of a sustainable food system and of a diverse and attractive rural environment.
- Provide critical components of a healthy diet for humans (even if imprudent use of animal products can result in health problems such as obesity, heart disease, cancer and infectious diseases).
- Play crucial roles in human culture and tradition, and through their interactions with people define human nature, provide companionship and contribute to aesthetic values.

It is also the case that, whoever (probably unwittingly) was responsible for it happening, humanity has now effectively assumed stewardship of the Earth's resources; and we are ethically required to act prudently to ensure that our descendents inherit a world worth inhabiting (Carpenter, 1998: 290). In summary, these considerations add up to a powerful endorsement of a *partnership* with animals, which will be to mutual benefit: one that replaces the currently dominant exploitative relationship. Moreover, there is a growing belief, doubtless informed by recent scientific advances, that farm animals are sentient beings who deserve respect for their intrinsic, as well as their instrumental, value (Mepham, 2000). Indeed, recent legislation in the EU (e.g. see Webster, 2006) and in Switzerland (Swiss Ethics Committee on Non-human Gene Technology, 2001) now requires that in our dealings with them due account be taken of animals' sentience and intrinsic worth.

While it seems clear that most intensive, and some extensive, forms of animal agriculture largely fail to respect adequately either animals' rights or welfare, one form of agriculture which could be claimed to do so is organic farming. So that if we are to devise an ethical approach to animal agriculture there is a *prima facie* case for saying that this is a sound starting point for our enquiries. For example, organic systems insist on low stocking densities, avoid any use of battery caging or similar confinement systems, and prohibit the use of antibiotics as growth promoters (Soil Association, 2000).

A key concept here is that of a notional *ethical contract* with animals. Social contract theory has a long history in political philosophy, but one of its more recent, and highly influential, advocates is the late John Rawls. For Rawls (1972: 3):

> A theory, however elegant and economical must be rejected if it is untrue: likewise, laws and institutions, no matter how efficient or well arranged, must be reformed or abolished if they are unjust.

And while justice is capable of different interpretations, Rawls (1972: 3) famously equated it with fairness, a definition he believed would be acceptable to 'free rational persons concerned to further their own interests'. Rawls proposed a conceptual device, which he called the 'veil of ignorance', to facilitate rational decision making in accordance with the principle of justice as fairness. With this hypothetical device in place, people were to act as though ignorant of their age, sex, wealth, intelligence, physical attributes etc. - and he envisaged that this would lead them to adopt two principles: first, equal liberties for all (allowing each as much liberty as is consistent with all others having the same) and, second, the difference principle (restricting social and economic inequalities to those that would benefit the least advantaged) (Rawls, 1972).

While Rawls himself excluded animals as ethical agents, because they are incapable of reasoning, others have argued that since some *people* may find themselves in this position (who are then referred to as 'moral patients', see Rodd, 1990: 241-242) there is a case for also including farm animals. If this were so, we should be faced with the task of devising an ethical contract that treated both people and animals fairly. Accepting that this is a *notional* contract (since animals clearly cannot 'sign up' to it), what it seems to require is that farm animals live better, or at the very least not worse, lives than they would otherwise have lived in the wild (Larrere and Larrere, 2000; Food Ethics Council, 2001; Lund *et al.*, 2004).

For some people, the flaw in the concept is that animals are almost inevitably deprived of their lives prematurely; hardly something a rational agent would agree to. Yet, to keep alive vast numbers of ageing, often sick animals would present almost insurmountable economic, humanitarian and logistical challenges. And, since in the wild older animals would usually die through predation, illness or exposure, domestication under the circumstances described may still be thought to constitute a more humane outcome. In fact, to be at all consistent, vegans are forced to adopt some arbitrary ethical positions. If the vegan's objective is to avoid animal suffering then there would seem to be no reason in principle to confine this to farm animals, or those animals otherwise under human care. But if individual wild animals were considered to have ethical standing, there would seem to be an ethical obligation to protect the prey from its predator, which would, even in the unlikely event that it could be achieved, inevitably have adverse effects on the well-being and survival of the predator. So even the most rigorous observance of veganism seemingly turns a blind eye to the extensive animal suffering that occurs in the wild.

Perhaps then an ethical position consistent with the 'facts of nature' must acknowledge that somewhat arbitrary dividing lines need to be drawn, and these are necessary even with respect to the animals in human care. Undoubtedly, however, the manner of death experienced by such animals - ideally, painless and unanticipated - is an important element in the notional contract. Currently, the rules for organic systems require no special provisions to ensure humane slaughter, above those applying to other animals.

Conclusions

Ethics is about deciding on the right actions, all things considered. Given the co-evolution of humans and non-human animals, and the vital roles those animals we have domesticated could play in a humane, well-managed, sustainable food system, there is a strong case for revising the basis of our continued relationship with them to one that is based on a notional contract: an arrangement by which animals continue to provide benefits to humans and the environment, but themselves live out better lives than they would in the wild. Attitudinal change could achieve much, but if such a contract is to be accorded appropriate weight it needs to be supported by: changes in legislation (e.g. entailing revision of WTO rules); amendments to animal welfare law; financial support for conversion to holistic, humane farming systems, such as organic farming; research into improved animal husbandry practices; and public education programmes which enlighten people at large on the needs, rights and benefits of animals as partners in a sustainable agriculture.

References

Bentham, J. (1948) [1823] *A Fragment on Government and Principles of Morals and Legislation.* Blackwell, Oxford, UK.

Budiansky, S. (1992) *The Covenant of the Wild.* Weidenfeld and Nicolson, London, UK.

Carpenter, S.R. (1998) Sustainability. In: Chadwick, R. (ed.) *Encyclopedia of Applied Ethics. Vol. 4* Academic Press, San Diego, USA, 275-293.

Dawkins, R. (1999) Foreword. In: Burley, J. (ed.) *The Genetic Revolution and Human Rights.* Oxford University Press, Oxford, UK, xv-xvi.

Department for International Development (2001) *Biodiversity: A Critical Issue for the World's Poor.* DFiD, London, UK.

FAO (2000) *World Watch List for Domestic Animal Diversity.* Third edition. FAO, Rome, Italy.

FAO (2005) *Ethical Issues in Fisheries.* FAO, Rome, Italy.

Food Ethics Council (2001) *Farming Animals for Food: Towards a Moral Menu.* FEC, Southwell, Notts, UK.

Francione, G.L. (2000) *Introduction to Animal Rights.* Temple University Press, Philadelphia, USA, 51-54.

Gold, M. (1999) Beyond the killing fields: working towards a vegetarian future. In: Tansey, G. and D'Silva, J. (eds) *The Meat Business.* Earthscan, London, UK, 169-183.

Hall, D.C., Ehui, S. and Delgado, C. (2004) The livestock revolution, food safety, and small-scale farmers: why they matter to us all. *Journal of Agricultural and Environmental Ethics* 17, 425-444.

Harrison, R. (1964) *Animal Machines.* Vincent Stuart, London, UK.

Hodges, J. (1999) Animals and values in society. *Livestock Research for Rural Development* 11(3) 1-6. [Accessed July 2007] Available from http://www.cipav.org.co/lrrd/lrrd11/3/hod113.htm

Hodges, J. (2005) Sustainable agriculture and food are at risk. *Livestock Production Science* 98, 225-230.

Ingold, T. (1994) Introduction. In: Ingold, T. (ed.) *What is an Animal?* Routledge, London, UK, 1-16.

International Livestock Research Institute (2004) *Achieving More with Less: Livestock as a Tool in Agricultural Intensification.* ILRI, Nairobi, Kenya.

Larrere, C. and Larrere, R. (2000) Animal rearing as a contract? *Journal of Agricultural and Environmental Ethics* 12, 51-58.

Locke, J. (1924) [1690] *Of Civil Government*. J.M. Dent, London, UK.

Lund, V., Anthony, R. and Rocklingsberg, H. (2004) The ethical contract as a tool in organic animal husbandry. *Journal of Agricultural and Environmental Ethics* 17, 23-49.

Mepham, B. (2000) Wurde der Kreature and the common morality. *Journal of Agricultural and Environmental Ethics* 13, 65-78.

Mepham, B. (2004) Animal diseases in context. *Journal of Agricultural and Environmental Ethics* 17, 331-340.

Mepham, B. (2005) *Bioethics: An Introduction for the Biosciences*. Oxford University Press, Oxford, UK.

Mepham, B. (2006) The ethical matrix as a decision-making tool, with specific reference to animal sentience. In: Turner, J. and D'Silva, J. (eds) *Animals, Ethics and Trade: The Challenge of Animal Sentience*. Earthscan, London, UK, 134-145.

Mill, J.S. (1910) [1863] *Utilitarianism, Liberty and Representative Government*. J.M. Dent, London, UK.

Rawls, J. (1972) *A Theory of Justice*. Oxford University Press, Oxford, UK.

Regan, T. (1985) The case for animal rights. In: Singer, P. (ed.) *In Defence of Animals*. Blackwell, Oxford, UK.

Regan, T. (1990) The Rights view. In: Anon. (ed.) *Animal-Human Relationships: Some Philosophers' Views*. RSPCA, Horsham, UK.

Rodd, R. (1990) *Biology, Ethics and Animals*. Clarendon, Oxford, UK.

Singer, P. (1990) The ethics of animal liberation: a summary statement. In: Anon. (ed.) *Animal-Human Relationships: Some Philosophers' Views*. RSPCA, Horsham, UK.

Soil Association (2000) *Livestock Standards: Incorporating EU Regulation 1804/1999*. Soil Association, Bristol, UK.

Swiss Ethics Committee on Non-Human Gene Technology (2001) *Die Würde des Tiers*, ECNH, Bern, Switzerland.

Webster, J. (2006) Ideals and realities: what do we owe to farm animals? In: Turner, J. and D'Silva, J. (eds) *Animals, Ethics and Trade: The Challenge of Animal Sentience*. Earthscan, London, UK, 149-158.

Williams, B. (1993) *Ethics and the Limits of Philosophy*. Fontana, London, UK.

Chapter 12

Beasts of a different burden: agricultural sustainability and farm animals

H. Buller and C. Morris

Introduction

> Arising from our animal natures, ethics needs no ground: but it runs aground in the conflict of our needs (Gray, 2002: 106)

Within both popular and policy imaginations, farm animals are central to the rural landscape as well as to the culture and identity of farming and rural places (Scruton, 1996). It might be assumed, therefore, that the lot of farm animals would be an important element within any discussion of agricultural sustainability and its associated policies and practices. Yet the relationship between farm animals and agricultural sustainability remains problematic and, at times, contradictory. For Korthals (2001: 814) 'the choice between environmentally sound agriculture and agriculture that treats animals acceptably is a difficult one'. This chapter seeks to interrogate the problematic nature of this relationship, principally as it is played out in a UK context. Specifically, the chapter investigates the varied assumptions about society-(farm) animal relations that lie within notions of agricultural sustainability. We ask whether the agricultural sustainability agenda serves the 'interests' of farm animals, provoking a different, more ethical, relationship between farm animals and society (than is the case in debates surrounding 'conventional' agriculture) or whether it merely serves to reinforce existing, and largely instrumental, animal–society relations, albeit locating them within a different moral framework. We identify two dominant discourses that link farm animals into debates on sustainability: the first of these sees farm animals as a *threat* to sustainability; the second sees animals as *vectors* for delivering sustainability. In recent years, these two conceptualizations have been joined by a new element in the incorporation of farm animals within conceptions of sustainability. This sees farm animals as the *targets* of sustainability, as notions of animal welfare become integrated into the very definitions of sustainable agriculture. We interrogate this third position, identifying some accommodations but also a number of contradictions inherent in its underlying assumptions about farm animal–society relations. In particular, reference is made to the implicitly anthropocentric nature of both agricultural sustainability and welfarism, to the tension between the individual animal subject (the focus of welfare) and animal collectives (the focus of environmental sustainability) and, to reprise Gray (2002), to the conflict of needs that the identification of animals as the targets of sustainability fails to resolve. In an attempt to move beyond some of these contradictions, the chapter ends with a discussion of an alternative way of thinking about farm animal–society relations that emphasises notions of relationality and co-constitution.

From the animal instrument to the animal subject

In early policy documents and scientific writing on sustainable development in the 1980s, farm animals, in and of themselves, are notable by their almost complete absence. This is also true, to an extent, of other animals, although both receive an implicit acknowledgement through discussion of the need to ensure 'biological diversity' (e.g. Brklacich *et al.*, 1991; Bowers, 1995; Cobb *et al.*, 1999). Where references can be found to farm animals within this literature they tend to be either cast as *threats* to or as *vehicles* for achieving sustainability. These two positions appear to be dominant until the late 1990s when they are joined by a third representation in which farm animal welfare becomes an important concern suggesting that farm animals have become a legitimate *target* of sustainability discourse.

Farm animals as a threat to agricultural sustainability

One of the earliest sustainability documents, the Brundtland Report (WCED, 1987), identifies domestic livestock, along with a burgeoning human population, as a threat to sustainable development. The report states that due to the increasing consumption of milk and meat since the 1950s, there are now globally 'about 1.4 billion cattle and buffaloes, 1.6 billion sheep and goats, 800 million pigs, and a great deal of poultry - all of which weigh more than the people on the planet' (WCED, 1987: 119-20; see also CIWF, 2004). Farm animals compete for space and resources with human populations, and are identified as elements of socially and environmentally unsustainable farming systems. With respect to the first of these misdemeanours it is argued that feeding grain to animals is less efficient than feeding it to humans directly; a significant nutritional consideration when addressing global problems of food shortage and starvation. The recently published report, *The Global Benefits of Eating Less Meat*, by the NGO Compassion in World Farming (CIWF), provides a contemporary illustration of this line of thinking:

> We are in the midst of what has become known as the 'livestock revolution' in the developing world. Meat consumption is projected to rise there by a massive 3% per annum until 2020, by which time 63% of global meat production and slightly more than half (50.3%) of global milk production are expected to be from the South. Rather than helping to alleviate world hunger, this 'revolution' is likely to prove detrimental to the task of feeding a growing human population ... This is because grain-fattened animals take more energy and protein from their feed than they return in the form of food for humans (CIWF, 2004: 22).

The key point here is that farm animals are being constructed as a threat to *human* welfare, or social sustainability, on a *global* scale. These anti-animal (or, at least, fewer animal) arguments are, more often than not, articulated in relation to debate surrounding global rather than national agricultural sustainability. In the UK, where the majority of farmed land is devoted to livestock production (either in terms of rearing the animals themselves or growing grain to feed them) and the livestock industry is relatively well organized politically, there have been few calls to reduce the number of farm animals. However, evidence can be found of both older and more current articulations of these ideas within the British literature. For example, in the shadow of the 'Club of Rome' in the early 1970s, Mellanby (1975) argued that a plant, rather than a livestock based, diet was the most sustainable way of feeding the growing British population. More recently, Bowler (2002), in his discussion of the principles of 'alternative agriculture', highlights the importance of a shift towards a national, crop-based (protein) diet so as to remove the inefficiencies associated with feeding cereals to livestock for intensive meat and milk production. This last example also points to the importance of the relative number (as distinct from their overall total number) of farm animals within particular farming systems. In short, highly industrialized, intensive, and concentrated livestock production systems are typically identified as the greatest threat to sustainability, particularly in the context of environmental sustainability.

The environment is therefore the second area in which farm animals have been cast as a threat to sustainability, again at a number of different scales; globally, nationally and more locally. Internationally, livestock are viewed as a significant contributor to the production of global warming gasses such as methane and of other atmospheric pollutants such as ammonia, nitrous oxide and carbon dioxide. It has been estimated that 10% of total global greenhouse gases are derived from animal manure (CIWF, 2004). More recently, the FAO has suggested that livestock currently produce more greenhouse gas emissions than cars (FAO, 2006). The relationship between loss of tropical rain forest and the creation of grain lands to feed cattle and/or for cattle ranching represents another cause for concern. Moreover, livestock production makes inefficient use of increasingly limited water resources. There is now widespread acceptance that water scarcity will become at least as important a constraint on future food production as lack of available land. While it takes only 500 l of water to produce 1 k of potatoes and 900 l to produce 1 k of wheat or alfalfa, it takes 350,000 and 100,000 l of water respectively to produce the same quantity of chicken or beef (Pimentel *et al.*, 1997). It has therefore been argued that 'by moving down the food chain … the same volume of water could feed two people instead of one' (Postel, quoted in CIWF, 2004: 25). In a UK context farm animals have also been cast as environmental threats, with the overstocking and overgrazing of pasture, and the pollution of water courses by intensive livestock systems (e.g. Lowe *et al.*, 1997; Evans, 2000) highlighted. In addition, farm animals have also been the cause of olfactory and aesthetic offence for many middle-class incomers into the countryside.

In short, when farm animals become too great in number overall (no matter what the scale of analysis or context), and too great in number in specific or localized contexts (as in the case of intensive systems of livestock production), they threaten the environmental and social sustainability of agricultural systems. Scale and context matter greatly in this debate because according to the scale of analysis farm animals can actually be a crucial medium or vector for realising improvements in sustainability. As Schiere and colleagues (2002: 150) observe:

> Changes in resource/demand patterns cause changes in the behaviour of livestock production systems. This implies that livestock can be essential for the sustainability of one system in one context and detrimental for the same or another system in a context elsewhere with other resource flows.

Farm animals as vehicles for agricultural sustainability

In contradistinction to the above, farm animals are identified as the means of achieving environmental, economic and social sustainability within agriculture. At least two dimensions to the inclusivity of farm animals within sustainability agendas can be identified. First, livestock are frequently key elements in the achievement and/or maintenance of environmentally sustainable farming systems usually conceived as extensive, grass-based livestock systems, where animal husbandry and pasturing practices are seen as vital to the maintenance of biodiversity and environmental quality. Much research has drawn an explicit link between extensive husbandry practices and the sort of farming/environment synergy that characterizes agricultural sustainability. This is evident within the context of agri-environmental policy. Take, for example, British Environmentally Sensitive Areas (introduced as a policy in 1986) which, in spite of their diversity in terms of landscape characteristics and habitats, all seek to maintain extensive livestock grazing and conversion from arable to grassland. The link is also made clear in work on high natural value farming systems (Hellegers and Godeschalk, 1998), upland farming (Bignal and McCracken, 1993), low intensity farming systems (Beaufroy *et al.*, 1994), and sensitive environmental area management (Evans, 2000). From a rather different perspective, research has suggested that a national diet based on large herbivores (mixed pasture and forage system) may have a less detrimental impact on field faunal biodiversity (e.g. field mice, voles and amphibians) than one based on an all-plant based agriculture (Davis, 2003). Likewise, other commentators have asserted the importance of mixed i.e. livestock *and* arable systems in realizing sustainability objectives (e.g. Pierce, 1993; Bowler, 2002; Schiere *et al.*, 2002; Lund and Olsson, 2006). Second, livestock based farming systems are often seen as critical to the maintenance of rural economies and, with them, the rural social fabric, particularly in marginal areas where arable farming is difficult, for example, in upland and alpine areas (IEEP *et al.*, 2004). In this way, farm animals have come to constitute a central pillar of agricultural sustainability. Accordingly, Tudge (quoted in Porritt, 2004: 5) argues that livestock 'hugely increase the overall economy of farming' when they have been raised 'according to the tenets of good husbandry (the ruminants to eat the grass on the hills and wet meadows, the pigs and poultry to clear up the leftovers)'. He goes on to conclude that the appropriate number of animals within agricultural systems 'judiciously deployed is more efficient, not less, than all-plant agriculture' (Tudge, *op cit.*).

In both of these contexts, farm animals constitute part of the means by which environmental and economic sustainability can be achieved. They deliver, through their managed agency, agricultural system efficiency, biodiversity, grassland maintenance and landscape value. Yet, the animals themselves are not explicitly identified as being the recipients of such discourses of agricultural sustainability. However, recent years have seen a subtle shift in the ways in which farm animals have been incorporated into discourses of sustainable agriculture.

The animal subject

While farm animals have become increasingly significant as key constituents within agricultural sustainability discourse, it is only very recently that the interests of farm animals themselves have become a *target* of sustainability discourse. This has occurred principally through the incorporation of farm animal welfare considerations as a distinct element within government and NGO definitions of agricultural sustainability, alongside the more 'established' dimensions of environment, economy and society (e.g. MAFF, 2000; Sustain, 2002). For example, in its pilot set of indicators for sustainable agriculture the UK Ministry of Agriculture, Food and Fisheries (MAFF, 2000: 5), now the Department for Environment, Food and Rural Affairs (Defra), identifies 'respecting a high level of animal welfare' as one of five components of sustainable farming. References to farm animal welfare can also found within the European Union's 1999 Rural Development Regulation (Buller and Morris, 2002), the UK Government's 1999 *Strategy for Sustainable Development*, the Policy Commission's report on *Farming and Food: a Sustainable Future* (2002) and in the UK Government's response to the Policy Commission, *The Strategy for Sustainable Food and Farming* (Defra, 2002a).

Within the scientific and policy universe that is focused on farm animal welfare (see Miele *et al.*, 2005), debate has also been moving towards and embracing a sustainability agenda, and in doing so is converging with, and reinforcing, the incorporation of animal welfare into the concerns, definitions and practices of the agricultural sustainability. Indeed it might be argued that these two previously distinct domains are becoming increasingly difficult to disentangle. Spedding (1994), for example, sees the adoption of animal welfare standards as central to the emergence of more sustainable agriculture, a position endorsed by the Farm Animal Welfare Council who maintain that 'farm animal welfare needs to be a prominent feature in assessing the sustainability of any system' (FAWC, 2002). Likewise, in describing a multidisciplinary tool for evaluating animal welfare in small ruminant breeding specifications, El Balaa and Marie (2006: 100) assert that this tool could be a 'basis for taking into account other issues associated with the sustainability of the production systems, such as environmental, social or economic aspects'. Animal welfare, therefore, is conceptualized as another element within sustainable production systems alongside the environment, society and economy. Defra's *Animal Health and Welfare Strategy* (AHWS) reinforces this view, in its statement that, 'the health and welfare of farmed animals makes a major contribution to the sustainability of the livestock sector, to the wider farming and food industry, and more broadly to the countryside, rural communities and the rural economy' (Defra, 2004: 13). The AHWS is significant as it sets out a framework and direction designed to maintain and improve animal health and welfare standards over the next decade (FAWC, 2005). The interweaving of a sustainability discourse into the AHWS provides further evidence of how sustainability has become agricultural policy orthodoxy.

Interrogating agricultural sustainability's targeting of animals

In this section of the chapter we examine more closely the recent targeting of farm animals within sustainability discourse, to explore the extent to which this involves a genuine reassessment of the relationship between farm animals and society. To do this, two 'ethical positionings' in this relationship are identified and discussed. The first of these identifies a natural philosophical correspondence between animal welfare and agricultural sustainability based on a shared anthropocentrism. Nevertheless, there is scope within both for an ethic of care, based on the dependence of farm animals on their human keepers. The second emerges from an acknowledgement of rights, which, as we show, becomes problematic within the context of environmental concerns with its emphasis on (ecological) collectives rather than individuals.

Anthropocentrism, humanism and the ethics of care and dependence

An initial compatibility between animal welfare and sustainability arises out of their shared assumptions about the relationship between animals and society and the ethic of care that underlies that relationship. Sustainability is based on a 'hierarchy which puts human beings above the natural world. Though some efforts have been made to introduce respect for nature into its definition … for the most part sustainable development remains anthropocentric. It is sustainability of human populations and their well-being which is at issue, rather than that of nature' (Dryzek 1997: 130; see also Dobson, 1996). Sustainability, therefore, is primarily, if not exclusively, concerned with human development and human inter-generationality; in which 'natural capital' is defined wholly in terms of its human relevance (Pearce, 1993). This hierarchy in natural relations is evinced in those livestock-based farming systems that are widely considered as key to achieving agricultural sustainability, but the individual animals upon which such systems are based are not the actual focus of concern. Instead, retaining their status as 'objects', they are regarded as the means to an end. Even where attempts have been made to place livestock at the centre of the analysis, as for example Yarwood and Evans (2000) do in their study of rare breeds of farm animals and the role of these in sustaining cultural landscapes, it is clear that ultimately the interest lies in what can be achieved for society rather than the animals themselves.

The evocation of farm animals as contributory elements or vehicles in the essentially human project of sustainability (using natural resources in a manner that does not deprive future human generations of their benefit) derives from, and indeed enshrines, the classic modernist position with respect to animals. In simple terms, such a position argues that animals are there for the benefit of humanity (Singer, 1975). This position has served humanity well though it has arguably had less beneficial effects for the animal 'kingdom' (Harrison, 1964). This suggests that sustainable development concerns may actually have little to offer farm animals themselves.

The incorporation of farm animal welfare within sustainability discourses arguably begins to challenge this modernist view of animals, suggesting a more critical, less human emphasis in agriculture that is moving us away from the exploitative nature of productivism. However, the drive towards the incorporation of animal welfare objectives into sustainability policy is by no means contradictory to human interests within the sustainability agenda. Indeed, one might argue that it further supports and reinforces this project as revealed in some of the UK Government's policy statements. Here, farm animal welfare is often represented as part of human 'social sustainability':

> Food and Farming also makes a broader social contribution in relation to public health, in particular through people's diet and through workplace safety, and in relation to animal welfare (Defra, 2002b: 7).

Similarly the Ministerial Foreword in Defra's AHWS highlights that 'disease "prevention [among farm animals] is better than cure"' and that this is 'fundamental to developing a sustainable food and farming industry' (2004: 5). This implies that although farm animals themselves are a focus of concern, through attention to their health and well-being, they are a means to an end, through their role in the creation of a sustainable food industry for the *people* working within it. Later, the document argues, 'Protection of public health is the paramount issue underlying animal health policy, but society also has real concerns about the treatment of animals and the protection of the environment' (Defra, 2004: 12).

The key point here is that while the current targeting of farm animals within sustainability discourse through welfarism might be interpreted as one element of a broader critique of productivism as modernism's highest point, sustainability itself has never been anti-modern, as evinced in the terminology of 'ecological modernisation'. Likewise, the new commitment to high animal welfare farming systems as a component of sustainable development has *not* in itself encompassed any shift in society's fundamental relationship to farm animals, that of 'objects' from which 'products' are derived thereby benefiting human society and the economy. Rather, as we have argued elsewhere (Buller and Morris, 2003) the welfarist approach, as it articulates with arguments and policies for sustainable farming, leads ultimately to a re-embedding of modernist, anthropocentric human–animal relationships in which animals are considered as different to, unequal to and dependent upon humans. This does not mean that they are denied, or are considered unworthy of, moral attention. On the contrary, the growth in popular concern for the welfare of farm animals is undeniable (Eurobarometer, 2005; Kjørstad and Kjaernes, 2006). Our point, however, is that this concern operates well within what are considered to be entirely acceptable (if often obfuscated) mechanisms of human domination (husbandry and slaughter). If we accept that farm animals are also distinct from (and unequal to) wild animals because they 'have been bred to docility, tractability, stupidity and dependency' (Callicott, 1992: 314), then our obligations to them are different to our obligations both to other humans and to wild animals. Our ethical relationship, derived from a growing acknowledgement of their dependency, becomes one of *care* (Adams and Donovan, 1996), through daily intervention, and utilitarianist concern for well-being and the reduction of suffering during that period (their entire lives) when we are responsible for them. They become, in Fellows' (2000: 588) words, our '*moral patients*'. Sustainable agriculture is not vegan but it can be caring.

Non-anthropocentrism and rights

Accepting farm animals as human chatels allows them to be comfortably positioned within a broad ethic of care that is, at one level, entirely compatible with sustainable agriculture's inherent anthropocentrism. However, although sustainability and welfarism are increasingly packaged together in a mutual agenda of responsibility for the non-human (however humanist this agenda might be), there remains nonetheless, a series of fairly fundamental differences between them, as a number of authors have pointed out. For Wenz (2001), there is a longstanding conflict between the individualistic concerns of welfarism and the more holistic approach of environmentalism (and, we might add, sustainability), a point also raised by Rawles (1997). She argues:

> Welfarists are concerned with individual, sentient animals. Any concern for habitats, ecological systems and processes is derivative, it is not for the habitat itself but only in so far as the habitat is necessary for the welfare of the sentient individuals within it. The primary concern of conservationists is with habitats, species, ecological systems and processes; in short with ecological collectives. Concern for individuals - sentient or otherwise - will be derivative (Rawles, 1997: 139).

While both might attribute intrinsic value to non-human animals, conservation and sustainability do so in terms of collective (ecological) value, as opposed to the individual value of embodied life that is the focus of welfare concern. In the former, we find a denial of animal individuality and hence subjectivity. In the latter, we find these very attributes providing the grounds for the accord of distinct moral status.

Hence deep ecology (one variety of what Dryzek (1997) calls 'green romanticism') calls for 'biocentric equality', where no species is regarded as more valuable or in any sense higher than any other species. Although deep ecologists recognize 'the full range of human interests in the nonhuman world as well as the interests of the nonhuman community' they:

> show no concern for animals once they are out of nature … Even animals 'in nature' are seen as fair game for deep ecologists, to be hunted and eaten as expression of the proper human place in ecosystems. For it is only organic wholes such as ecosystems that are to be preserved, not individual creatures (Dryzek, 1997: 157, 158).

Likewise, animal liberation (for Dryzek (1997), a form of 'green rationalism') 'somehow fits uneasily in green discourse because it is weakly ecological, some would say anti-ecological. For in its concern with individual creatures, it can lose sight of larger ecological connections' (Dryzek, 1997: 184). While animal welfare concerns seek to reduce the suffering of animals, and in particular that suffering which results from human action, environmental conservation (one important dimension of agricultural sustainability) has not had such an explicit objective. Indeed, human interference in 'natural' suffering is generally regarded as an anathema, both morally and managerially (Bovenkerk *et al.*, 2003). In certain situations, these two perspectives may find themselves in direct opposition. Indeed, such was the subject of the often violent debate between Baird Callicott (1980), for whom domesticated animals were so humanized and manufactured as to be inherently incapable of being 'liberated', and Tom Regan (1983), for whom the rights of any individual animal, especially farm animals, were, by definition, greater than any Leopoldian 'rights' of the ecosystem. Examples abound of practical conflicts between the two positions; for example domesticated animals bred for and fed to zoo animals protected for their rarity, 'pest species' destroyed for environmental reasons and so on.

However, for many writers, a truly liberationist stand provides a challenge not only to environmental conservation and sustainability (Sagoff, 1984) but also to the very animals it seeks to 'liberate'. As the Food Ethics Council (2001: 6) has recently suggested: 'It would be a totally perverse act, resulting from a misguided sense of compassion, to attempt to return such domesticated animals to "the wild", even assuming such territory could be found. They simply could not survive' (see also Budiansky, 1992). Major ecosystem change would accompany the 'liberation' of farm animals into the 'wild', grassland ecologies and the different animal species that have adapted to them would alter significantly while the farm animals themselves, deprived of the support and human intervention on which they had become dependent, would die in their millions.

The moral accomodation that we identified in the previous section, therefore, begins to break down and animal welfare positions are revealed as being less immediately compatible with the arguments of sustainable agriculture. If we recognize animal subjectivity and individuality, then the ethic of care resulting from an ackowledgement of dependency is arguably no longer a sufficient basis for our engagement. Indeed, as Jamieson (1998) points out, we enter, in many ways a more conventional ethical sphere of justice, moral obligations and rights that extend a long way beyond the moral codes that make deliberate harming of animals, as inferior and subordinate, merely unacceptable, except within certain contexts, notably agriculture. Moral unacceptability is exclusively for humans. Rights, argue animal rights activists such as Regan (1983), necessarily include the living non-human and put them on, if not an equal (Regan, 1983), then at least a comparable (Singer, 1990) footing with human animals. At one level, such a position offers a fundamental challenge to the husbandry and slaughter of individual animals. At another level, it contests the de-subjectivity of individual animals for some greater 'ecological' good. Both offer a robust critique of the apparent collusion between sustainability and welfare.

Beyond sustainability

Fundamental to the concept of sustainability is a sense of continuity and protection. What we want to argue in this final section is that new ways of thinking about the relationship of humans to non-humans, that are emerging from contemporay social science, are challenging the continuity of the hitherto dualistic assumptions of that relationship particularly as they relate to the role and place of animals within sustainability debates. Here we discuss a different kind of ethical entanglement between human animals and non-human animals that points to an alternative (more enlightened) conceptualization of the relationship between animal welfare and sustainability, one based upon a holistic and shared sense of co-constitution, relationality and mutuality and which draws upon notions of 'hybrid communities', as developed by Lestel (2004) and others. Although welfare advocates (and liberationists) are keen to accord to animals the same, or similar rights as they would enjoy as humans, some categorical division of animal and human nonetheless persists (and indeed, in many contexts, needs to persist, relationality should not be a negation of either animality or humanity). Moreover, it entirely categorizes the issue of rights, representation and who, or what, is going to do anything about them. Unless animals are going to be able to do something about their lot on their own (à la *Animal Farm*) a degree of human moral appropriationism and surrogate agency becomes indispensible (Lynn, 1998). However, in the drive to creating a sustainable world, farm animals are no longer merely the automata, the beasts of burden and 'meat on legs' of mass productionism. Nor are they the 'blank sheets of paper' of our self-invested power of naming. Instead, the multifunctional cow that gives birth to beef calves, that feeds them and that ends up as meat also provides a mobile landscape feature, a grass nurturing and manuring machine, a contact point for non-human nature to those wandering the rural landscape, a material agent in the various networks that make places and an undeniable component of human performance within those places. As such, farm animals within sustainability assume a complex array of meanings, representations, functions and agencies that intertwines with human beings in a number of complex material and affective ways. The animal presence in human/non-human social life is, for Whatmore (2002: 32) 'corporeal, creative and consequential'. Farm animals have been re-constructed (and us along with them), as the Sustainable Development Commission (2003) suggests: 'animal health and welfare cannot easily be separated from the complexities of the real world, as they are intrinsically tied up with both human and environmental health and welfare'. It is this 'intrinsically tied up' notion that suggests new ways of conceptualizing the relationship between animal welfare and environmental welfare (as an element of agricultural sustainability). Whatmore (2003:

26) sums it up in asserting that the 'categorical distinctions between humans, animals and machines are seen to be uncertain as their properties are blurred by a proliferation of hybrid entities and networks'. As we have said above, farm animals already occupy that hybrid terrain both metaphorically (viz; Callicot's (1980) infamous likening of a flock of sheep to a four-wheel drive vehicle) and materially (e.g. Dolly the sheep). They are, in many ways, different from wild animals, yet they are equally far closer to wild animals than they are to human animals. They are highly dependent upon humans; they have, as species, been essentially 'made' by humans even though, as individuals, they are clearly also 'made' by other animals and can die of ('natural') causes other than their slaughter for meat. If to be 'natural' brings with it a certain moral worth, does their artificiality deny them such worth? Their position within the sustainability agenda occasions them to cross and co-exist within these 'subject'/'object', wild/domesticated, dependent/independent, individual animal/whole species divides (or occasionally to remain enclosed within them) in accordance with the various roles they are required to play under the conflicting needs of human ethical concerns. In terms of the current argument, what is needed is a more relational account of sustainability, in which farm animals and humans form a broader collective of shared interests and, consequently, a new framework for the ethical accountability of farm animals (and other animals) within such a relational sensibility.

The first is not so difficult to identify. The interests of farm animals are significantly threatened by unsustainable practices, agricultural or otherwise. Farm animals occupy, impact upon and are affected by the environment. Hedgerow removal deprives them of fodder and shelter against sun, wind and rain; climate change affects their long term survival as much as it does our own; environmental pollution will harm them just as it does humans. The suggestion that farm animals are merely the 'vehicles' of sustainability does not give enough weight to the notion that they are also its beneficiaries as members of what Midgley (1995) amongst others, calls the 'mixed community' in which humans and animals are intertwined (Noske, 1996). The second is more complex but builds from the first for, critically, these mixed or hybrid communities are constituted of far more than merely instrumental and functional relations. They are, or can be, representative and symbolic, reciprocal and affective iterative networks of human/non-human, social/natural relations and assemblages of human and non-human 'persons' (Lestel, 2004) and existences (Descola, 2005). From these articulations that blur and transcend the more familiar boundaries and thereby stretch beyond the 'anthropological machine' of dualistic Western thinking (Agamben, 2004), a new ethical relationship is required, based not on difference, but upon collective ethical community (Latour, 2004). Going beyond the mere reconciliation of welfare and sustainability, thereby demands a repositioning and inclusion of the (farm) animal as co-member of that community with the attendant moral consideration that this implies. Recognizing the key role that farm animals play, even a repentant Callicott (1988: 167) acknowledges that 'barnyard animals, over hundreds of generations, have been genetically engineered ... to play certain roles in the mixed community ... To condemn the morality of these roles ... is to condemn the very being of these creatures'. Yet these roles are still defined in instrumental and functional terms. As Lestel (2004) points out, it is paradoxical that our close proximity and our established, yet essentially functional, relations with respect to domesticated animals - key manoeuvres in the construction of community - have also permitted their greater (though far from universal) subjectivization. Acknowledgement of that subjectivity, recognition of individual animal histories and their intertwining relationships with our own come to define the moral realm (Clement, 2003). And relationships are affective. For feminist writers such as

Clement (1996) and Davis (1995), environmentalism and the arguments of sustainability have tended to steer away from the affective and 'feeling-based' approaches to ethics in favour of the harsher, de-individualized and more rational notion of justice. Yet recognition of animal sentience and 'feelings' (and particularly farm animal 'feelings') has been long in coming, frequently dismissed either as non-science or as simply un-testable and thereby un-knowable (Dawkins, 2005). A more relational, less 'human' sustainability whose 'target' is the hybrid communities of human and non-human seems to us an entirely necessary way forward. As Cary Wolfe (2003: 207) puts it: 'the only way to the "there" in which animals reside is to find them "here" in us and of us, as part of a plurality for which perhaps even "the animals" in the plural, is far too lame a word'.

Conclusion

In this chapter we have examined the debate surrounding agricultural sustainability for the ways in which it constructs, interprets, discusses and analyses the relationship between society and farm animals. In particular, we have been concerned to highlight the inconsistencies and contradictions in the far from simple task of bringing together an animal welfarist and sustainability agenda. Crucially, though, the shift from representing animals as threats to sustainability to the current position where animal welfare is now commonly included within discourses of agricultural sustainability has not fundamentally reconfigured the relationship between human and in-human worlds. However, it has brought about some sort of repositioning in this relationship. It does this, first, by making animals visible again thereby exposing and deconstructing the relations that we create for them (Baker, 2001). The incorporation of farm animals within sustainability debates may help to generate greater sympathies amongst the human population towards farm animals. With sustainability now a much more prominent notion within policy and public debates, including farm animals within discussion about sustainability will further increase the attention paid to their needs, rights and welfare. Expanding the circle of human concern to incorporate animals is something that Murdoch (2003), following Singer (1975) and others, has recently called for. Second, it allows humans, albeit in an indirect manner, to give animals a voice. Giving animals a voice can be extended to reinterpreting rural spaces, and their sustainability, in terms of animalian agency and intentionality, rather than exclusively human agency and intentionality. As Budiansky (1994) shows, animal domestication is not an entirely one-sided affair, but rather a co-evolved relationship. Jones (2003: 300), referring to Ingold (1994), writes that 'animals … build and dwell in the world in creative ways which are outside human enrolment and often in conflict with its networks'. Farm animals thereby become part of the sustainability project, not for the way they are managed in groups and herds by humans, but as a result of the relationality between their own individual intentions, behaviour, agency and use of space and nature (however limited these might be) and those of humans. Agricultural sustainability can be conceived therefore as a 'more than' both human and animal project, the collective endeavour of a relational community.

References

Adams, C. and Donovan, J. (eds) (1996) *Beyond Animal Rights: A Feminist Caring Ethic for the Treatment of Animals*. Continuum, New York, USA.

Agamben, G. (2004) *The Open: Man and Animal*. Stanford University Press, Stanford, USA.

Baker, S. (2001) Animals, representation, and reality. Guest Editor's Introduction. *Society and Animals* 9(3) 189-201.

Beaufroy, G., Baldock, D. and Clark, J. (eds) (1994) *The Nature of Farming: Low Intensity Farming Systems in Nine European Countries*. IEEP, London, UK.

Bignal, E. and McCracken, D. (1993) Nature conservation and pastoral farming in the British uplands. *British Wildlife* 4, 367-376.

Bovenkerk, B., Stafleu, F., Tramper, R., Vorstenbosch, J. and Brom, F.W.A. (2003) To act or not to act? Sheltering animals from the wild: a pluralistic account of a conflict between animal and environmental ethics. *Ethics, Place and Environment* 6(1) 13-26.

Bowers, J. (1995) Sustainability, agriculture and agricultural policy. *Environment and Planning A* 27, 1231-1243.

Bowler, I. (2002) Developing sustainable agriculture. *Geography* 87(3) 205-212.

Brklacich, M., Bryant, C. and Smit, B. (1991) Review and appraisal of concept of sustainable food production systems. *Environmental Management* 15(1) 1-14.

Budiansky, S. (1992) *The Covenant of the Wild*. Harvard University Press, New Haven, USA.

Budiansky, S. (1994) A special relationship: the co-evolution of human beings and domestic animals. *Journal of the American Veterinary Medicine Association* 204, 365-368.

Buller, H. and Morris, C. (2002) *Farm Animal Welfare Aspects of the Common Agricultural Policy. Report to the RSPCA*. CCRU, University of Gloucestershire, Cheltenham, UK.

Buller, H. and Morris, C. (2003) Farm animal welfare: a new repertoire of nature-society relations or modernism re-embedded? *Sociologia Ruralis* 43(3) 216-237.

Callicott, J.B. (1980) Animal ethics: a triangular affair. *Environmental Ethics* 2, 311-338.

Callicott, J.B. (1988) Animal liberation and environmental ethics: back together again. *Between the Species* 4, 163-169.

Callicott, J.B. (1992) The search for an environmental ethic. In: Regan, T. (ed.) *Matters of Life and Death*. Third Edition, McGraw-Hill, New York, USA, 322-382.

Clement, G. (1996) *Care, Autonomy and Justice*. Westview Press, Boulder, USA.

Clement, G. (2003) The ethic of care and the problem of wild animals. *Between the Species. An Electronic Journal for the Study of Philosophy and Animals,* III. [Accessed July 2006] Available from http://cla.calpoly.edu/~jlynch/issueIII.html

Cobb, D., Dolman, P. and O'Riordan, T. (1999) Interpretations of sustainable agriculture in the UK. *Progress in Human Geography* 23(2) 209-235.

Compassion in World Farming Trust (2004) *The Global Benefits of Eating Less Meat*. Report for CIWF Trust by Mark Gold, CIWF Hampshire, UK.

Davis, K. (1995) Thinking like a chicken: farm animals and the feminist connection. In: Adams, C. and Donovan, J. (eds) *Animals and Women: Feminist Theoretical Explanations*. Duke University Press, Durham, North Carolina, USA, 192-214.

Davis, S.L. (2003) The least harm principle may require that humans consume a diet containing large herbivores, not a vegan diet. *Journal of Agricultural and Environmental Ethics* 16, 387-394.

Dawkins, M. (2005) Through animal eyes. Paper to the CIWF Conference, *From Darwin to Dawkins: the Science and Implications of Animal Sentience*. CIWF, London, UK.

Defra (2002a) *The Strategy for Sustainable Farming and Food: Facing the Future*. HMSO, London, UK.

Defra (2002b) *Farming and Food's Contribution to Sustainable Development: The Current Situation and Future Prospects*. HMSO, London, UK.

Defra (2004) *Animal Health and Welfare Strategy for Great Britain*. HMSO, London, UK.

Descola, P. (2005) *Par-delà Nature et Culture*. Gallimard, Paris, France.

Dobson, A. (1996) Environmental sustainabilities: an analysis and a typology. *Environmental Politics* 5(3) 401-428.

Dryzek, J. (1997) *The Politics of the Earth: Environmental Discourses.* Oxford University Press, Oxford, UK.

El Balaa, R. and Marie, M. (2006) Animal welfare considerations in small ruminant breeding specifications. *Journal of Agricultural and Environmental Ethics* 19(1) 91-102.

Eurobarometer (2005) Attitudes of consumers towards the welfare of farmed animals. *Eurobarometer,* 229/Wave 63.2. Eurobarometer, Brussels, Belgium.

Evans, N. (2000) The impact of BSE in cattle on high nature value conservation sites in England. In: Millward, H., Beesley, K., Ilbery, B. and Harrington, L. (eds) *Agricultural and Environmental Sustainability in the New Countryside.* Nova Scotia Agricultural College, Truro, Nova Scotia, Canada, 92-110.

FAO (2006) *Livestock's Long Shadow.* FAO, Rome, Italy.

FAWC (2002) *FAWC's Response of 2 September 2002, to Defra's Consultation on the New Priorities Board for Publicly Funded Research.* Press Release. FAWC, London, UK.

FAWC (2005) *Welfare Implications of Farm Assurance Schemes.* FAWC, London, UK.

Fellows, R. (2000) Animal belief. *Philosophy* 75, 587-598.

Food Ethics Council (2001) *Farming Animals for Food: Towards a Moral Menu.* Food Ethics Council, London, UK.

Gray, J. (2002) *Straw Dogs.* Granta Books, Cambridge, UK.

Harrison, R. (1964) *Animal Machines.* Vincent Stuart, London, UK.

Hellegers, P. and Godeschalk, F. (1998) *Farming in High Nature Value Regions; The Role of Agricultural Policy in Maintaining HNV Farming Systems in Europe.* Agricultural Economics Research Institute, The Hague, The Netherlands.

IEEP, Land Use Consultants and GHK Consulting (2004) *An Assessment of the Impacts of Hill Farming in England on the Economic, Environmental and Social Sustainability of the Uplands and More Widely.* Report to Defra, IEEP, London, UK.

Ingold, T. (ed.) (1994) *What is an Animal?* Routledge, London, UK.

Jamieson, D. (1998) Animal liberation is an environmental ethic. *Environmental Values* 7, 41-57.

Jones, O. (2003) The restraint of beasts: rurality, animality, Actor Network Theory and dwelling. In: Cloke, P. (ed.) *Country Visions.* Pearson, London, UK, 283-237.

Korthals, M. (2001) Ethical dilemmas in sustainable agriculture. *International Journal of Food Science and Technology* 36, 813-820.

Kjørstad, I. and Kjærnes, U. (2006) *Consumers and Animal Welfare: A Comparative European Analysis.* The National Institute for Consumer Research, Oslo, Norway. [Accessed Dec 2006] Available from http://www.welfarequality.net /everyone/29696

Latour, B. (2004) *Reassembling the Social.* Clarendon Press, Oxford, UK.

Lestel, D. (2004) *L'Animal Singulier.* Seuil, Paris, France.

Lowe, P., Clark, J., Seymour, S. and Ward, N. (1997) *Moralising the Environment: Countryside Change, Farming and Pollution.* UCL Press, London, UK.

Lund, V. and Olsson, A.S. (2006) Animal agriculture: symbiosis, culture, or ethical conflict? *Journal of Agricultural and Environmental Ethics* 19, 47-56.

Lynn, W. (1998) Animals, ethics and geography. In: Wolch, J. and Emel, J. (eds) *Animal Geographies.* Verso, London, UK, 280-297.

MAFF (2000) *Towards Sustainable Agriculture: a Pilot Set of Indicators.* MAFF, London, UK.

Mellanby, K. (1975) *Can Britain Feed Itself?* Merlin Press, London, UK.

Midgley, M. (1995) The mixed community. In: Sterba, J. (ed.) *Earth Ethics.* Prentice Hall, Englewood Cliffs, New Jersey, USA, 80-90.

Miele, M., Murdoch, J. and Roe, E. (2005) Animals and ambivalence: governing farm animal welfare in the European food sector. In: Higgins, V. and Lawrence, G. (eds) *Agricultural Governance: Globalization and the New Politics of Regulation.* Routledge, London, UK, 169-185.

Murdoch, J. (2003) Geography's circle of concern. Editorial. *Geoforum* 34(3) 287-289.

Noske, B. (1996) *Beyond Boundaries.* Black Rose Books, Montreal, Canada.

Pearce, D. (1993) *Blueprint 3: Measuring Sustainable Development.* Earthscan, London, UK.

Pierce, J. T. (1993) Agriculture, sustainability and the imperatives of policy reform. *Geoforum* 24(4) 381-396.
Pimentel, D., Houser, J., Preiss, E., White, O., Fang, H., Mesnick, L., Barsky, T., Tariche, S., Schreck, J. and Alpert, S. (1997) Water resources: agriculture, the environment and society. *Bioscience* 47(2) 97-106.
Policy Commission on the Future of Farming and Food (2002) *Farming and Food: a Sustainable Future.* Cabinet Office, London, UK.
Porritt, J. (2004) Foreword. In: CIWF Trust, *The Global Benefits of Eating Less Meat*. CIWF Trust, Petersfield, UK, 4-7.
Rawles, K. (1997) Conservation and animal welfare. In: Chapell, T. (ed.) *The Philosophy of the Environment*. Edinburgh University Press, Edinburgh, Scotland, 135-155.
Regan, T. (1983) *The Case for Animal Rights*. Routledge, London, UK.
Sagoff, M. (1984) Animal liberation and environmental ethics: bad marriage, quick divorce. *Osgoode Hall Law Journal* 22, 297-307.
Schiere, J., Ibrahim, M., and van Keulen, H. (2002) The role of livestock for sustainability in mixed farming: criteria and scenario studies under varying resource allocation. *Agriculture, Ecosystems and Environment* 90, 139-152.
Singer, P. (1975) *Animal Liberation*. New York Review of Books, Random House, New York, USA.
Singer, P. (1990) *Animal Liberation.* Second edition. Random House, New York, USA.
Scruton, R. (1996) *Animal Rights and Wrongs*. Demos, London, UK.
Spedding, C. (1994) Animal in welfare. *Journal of the American Veterinary Medical Association* 204(33) 384-387.
Sustain (2002) *Sustainable Food Chains. Briefing Paper 1: Local Food; Benefits, Obstacles, Opportunities*. Sustain, London, UK.
Sustainable Development Commission (2003) *Preparing an Animal Health and Welfare Strategy for Great Britain.* [Accessed June 2004] Available from http://www.sustainable-development.gov.uk/forum/03.htm
UK Government (1999) *A Better Quality of Life: A Strategy for Sustainable Development for the UK.* CM4345, HMSO, London, UK.
Wenz, P. (2001) *Environmental Ethics Today*. Oxford University Press, Oxford, UK.
Whatmore, S. (2002) *Hybrid Geographies. Natures Cultures Spaces*. Sage, London, UK.
Whatmore, S. (2003) Animal geographies. In: Johnston, R.J., Gregory, D., Pratt, G. and Watts, M. (eds) *The Dictionary of Human Geography*. Fourth edition. Arnold, London, UK, 25-26.
Wolfe, C. (2003) *Animal Rites*. University of Chicago Press, Chicago, USA.
WCED (1987) *Our Common Future*. Oxford University Press, Oxford, UK.
Yarwood, R. and Evans, N. (2000) Taking stock of farm animals and rurality. In: Philo, C. and Wilbert, C. (eds) *Animal Spaces, Beastly Places*. Routledge, London, UK, 98-114.

Chapter 13

Agricultural biotechnology and ethics: for or against nature?

B.R. Johnson

Introduction

To explore moral and ethical issues relevant to agricultural biotechnology it is worth first examining those dimensions of agriculture and selective breeding. Humans are part of the natural fauna of the Earth. This might seem to be an obvious statement but it is a fact often overlooked in debates about the morality of human impacts on the planet and ethical approaches to producing what we need to survive. In common with many other species, we modify our habitat to suit our needs. Even lowly creatures such as some species of ants select and nurture crops, but these are fungi, not the wheat, maize, rice, fruit and vegetables we cultivate for our survival. Just like ants who 'herd' aphids for their sweet secretions, we ranch and farm many kinds of animals, not only for food but also for hides and bones that we use for many different purposes.

Most of the organisms humans domesticate are far from natural; they have been selected from mutations and hybrids over many generations of breeding, usually carried out by farmers. Sometimes crop and livestock breeding was deliberate but it often took the form of farmers recognizing when a natural event produced an outstandingly useful variety. For example, primitive cultivated wheat crossed naturally with wild wheat and the additional genetic information turned a rather mundane cereal into a prolific and robust crop upon which whole civilizations were founded. As Darwin (1859: 8) said:

> No case is on record of a variable being ceasing to be variable under cultivation. Our oldest cultivated plants, such as wheat, still often yield new varieties; our oldest domesticated animals are still capable of rapid improvement and modification.

With our vastly increased knowledge of plant and animal genomes and the genetic structure of populations, Darwin's observation is even more relevant today.

In the past there have been few disputes over the morality of genetically modifying natural organisms through selective breeding, and there have been no explicit ethical codes (other than those concerned with animal welfare) to provide a framework within which plant and animal breeders operate. This is in contrast to other areas of scientific endeavour such as medicine and human genetics, where rigorous ethical codes have been developed to govern research and development worldwide (for examples see Jackson, 2006; NIREHG, 2007). Where plant breeding is concerned ethics has not been an issue as plants are quite clearly non-sentient organisms and until recently the genetic make-up of crops has not been seen as an issue either morally or ethically. In ornamental horticulture, genetic freaks are highly prized and much admired. They are the roses, begonias and dahlias seen at every flower show, and the vegetables seen at every table. Animal breeding has also until fairly recently not been considered by the public to be an area of moral or ethical concern. The production of a wide range of genetic curiosities such as pet dogs and birds has been acceptable to most people, although these organisms can have gross phenotypic distortions that make them unable to survive in the wild and can in some dog and cat breeds cause them considerable distress and discomfort. Moral and ethical concern has centred on the conditions in which pet, farmed and experimental animals are kept, rather than their genetic composition and phenotypic appearance.

Whilst over the past 5000 years humans have modified other organisms to suit our purposes, of all species we have had the greatest impact on the Earth's habitats. Not only have we destroyed vast areas of natural forest, marshes and steppe, replacing them with arable cultivation, but we have also used grazing animals and intensive management to produce highly modified grasslands on which livestock production depends (Conner *et al.*, 1998; Kristensen *et al.*, 2005). There is very little that is 'natural' about farmed landscapes, although some natural species adapt to, and co-evolve with, arable farmland and grasslands to produce pseudo-ecosystems that are now the only refuge for some species of plants and animals. This gross destruction and modification of the Earth's ecosystems has been an integral part of human society's growth and development and much of it is necessary for our continued survival and well-being, but it is far from sustainable because it has often failed to continue to provide food in the long term.

The vast deserts of North Africa and some parts of Asia are partly produced by humans; the result of hundreds of years of overgrazing and inappropriate arable cultivation coupled with over-abstraction of natural water resources. Not only did our ancestors ruin productive farmland, they also brought about the downfall of their own civilizations, such as those in the Negev region and in some parts of South America. Modern unsustainable agriculture relying on 'fossil' aquifers is currently being practised in the Mediterranean region and will probably have the same effect of desertification as irrigated systems did in Libya centuries ago (Mattingly *et al.*, 2003). We are also now extending unsustainable management systems from terrestrial to coastal and marine environments where shellfish and fish farming not only degrade their surroundings but also have damaging impacts on marine fish populations such as sand eels that are harvested to feed the farmed fish (RSPB, 2004). Such destruction continues apace, exacerbated by global warming and increases in both human population numbers and wealth, with soil degradation and destroyed natural habitats as much a feature of the 20th century as in earlier times.

The use of transgenic technology, producing GM organisms with novel (and sometimes radical) characteristics, has the potential to transform agriculture and other human activities such as forestry and aquaculture. New crops with drought and salinity tolerance, pest and disease resistance and tolerance to herbicides have already been produced (Johnson, 2003; FAO, 2005). These crops can enable changes in cultivation, forestry and aquacultural systems that could be part of a more sustainable future, yet there are deep concerns within civil society over the wisdom of using this technology and its products. Ironically, these concerns are partly fuelled by a mistrust of 'conventional' agriculture following outbreaks of BSE and foot and mouth disease.

The morality and ethics of agriculture

It is worthwhile examining the moral justification of modifying organisms and habitats for our own use and whether we have ethical frameworks for these activities. At one level, the utilitarian view is that humans are morally justified in exploiting nature for our ends because without doing so we would probably either become extinct or be forced to live in primitive groups with little 'civilisation'. Others, such as Meadows and colleagues (1972) and Daily and Ehrlich (1992), argue that with unbridled use of new technologies we have taken exploitation to industrial levels that now threaten the Earth's natural systems to such an extent that humans may not survive into the future, resulting in the same outcome as the first argument. Heidegger (1977) argued in his polemic on the industrialization of agriculture that humans are intrinsically morally justified in exploiting nature but must do so only if they do not destroy the resources that continue to provide sustenance. Heidegger believed that we have a moral right to exploit natural resources only when we have a deep understanding of the consequences of doing so, an understanding engrained in the traditional knowledge of peasant farmers. For Heidegger the moral issue here was not whether or not we should exploit nature - it is clear that we should - but that we should do so in a way that is sustainable. There is a powerful argument that we have a moral duty to hand on to successive generations the same levels of resources that we have enjoyed. This moral imperative is embodied in the 1987 Brundtland definition of sustainable development (WCED, 1987: 43) as 'development that meets the needs of the present generation without compromising the ability of future generations to meet their own needs'. It was further developed by the Rio Declaration of 1992 (UNEP, 1992) where for the first time most nations subscribed to the concept of sustainable development in terms of the conservation and sustainable use of all the Earth's resources, including biodiversity and genetic resources. We still have a long way to go in translating those fine principles into political and economic reality but the Rio Declaration was a milestone in consensus morality. Sustainability, perhaps a new moral, lies at the heart of growing concerns in the 20th century about population growth, destruction of biodiversity and global warming.

Translating moral values, both intrinsic and extrinsic, into practical use often leads to the production of ethical codes. In medicine and human genetics for example, ethical frameworks for research, development and treatments are highly developed and subject to constant scrutiny and review by scientists, politicians and civil society, not only because moral values are dynamic, changing with the ebb and flow of religion and secularism for example, but also because technologies develop very rapidly, giving rise to new moral and ethical issues (Jackson, 2006). The discoveries of gene therapy and mammalian cloning are two recent examples of where ethical codes have had to be formulated urgently because the science outpaces our ability to respond to the issues raised (for discussion of gene therapy see Walters and Palmer, 1996 and for cloning issues see Brannigan, 2001). Perhaps the most surprising feature of ethical codes in these areas is that they are widely welcomed by industry, mainly because they set common boundaries for research and development but also because if they are widely agreed by civil society industry can be reasonably confident that products from these new technologies will be acceptable to customers (for an example in biotechnology and discussion see Lehmann, 1997). There are few ethical codes governing agricultural research and development, but some countries apply codes to the way agriculture is carried out, an early example being the 1985 *Code of Good Agricultural Practice* for England and Wales developed by MAFF in relation to pollution legislation (Lowe *et al.*, 1997). Those codes concerned with farm animal welfare (e.g. *Animal Welfare*

Act, 2006; Defra, 2006a) have been developed in much the same way as those in human medicine and genetics, with wide participation including the farming industry, NGOs and politicians. Significantly there appear to be no such ethical codes developed for crop and livestock breeding or for the development of novel farming systems, despite the controversy that has arisen from the use of transgenic biotechnology in these areas. Perhaps the only example of an ethical code for farming systems is that governing organic farming (Defra, 2005), developed in the UK by the Soil Association, supported by the UK government, and later adopted worldwide.

Are moral values and ethical codes relevant to crop and livestock breeding?

In some parts of the world and in some religions there have been traditional moral objections to the farming and consumption of certain animals, such as pigs in Islamic and Jewish cultures and dogs and cats in many 'western' cultures, possibly based on ancient knowledge of the high risks of contracting diseases and internal parasites, coupled with the bonds established between humans and their pets. This aversion to harming animals kept as pets has not however prevented the production of genetic 'freaks' in pet animals, some of which (such as 'roller' and 'tumbler' pigeons, extreme forms of chickens, and the wide range of pet dogs) are seen as highly prestigious in some cultures. Many important crops are also what Amman (personal communication, 2000) rightly describes as 'genetic monsters', including maize, wheat and many fruits and vegetables. Until very recently the new genetic combinations required to produce novel crops and livestock have come from mutation, recombination and hybridization - processes that have been seen as 'natural' despite a high degree of human intervention such as the use of mutagens, embryo rescue and cell fusion to produce exotic varieties and hybrids (Allard, 1999). Farmers and consumers, even those who prefer to grow and consume organic food, see the products of these breeding programmes as acceptable even though we have little knowledge of the genetic changes that have taken place. Even organic farmers use plant varieties produced by artificial mutagenesis, although some organic farming organizations have reservations about such techniques (van Bueren *et al.*, 1998).

Other than animal welfare, there has been little consideration of the moral and ethical dimensions of the impacts of industrialized and intensive small plot farming on the functionality and aesthetic values of ecosystems and landscapes, although there has been a strong academic critique of the ecological and polluting impacts of such farming (e.g. Carson, 1962; Shoard, 1980; Body, 1982; Conway and Pretty, 1991). In the latter half of the 20th century the morality of destroying vast areas of rainforest and semi-natural grassland in order to convert them to agriculture has been called into question, for example, by the UN World Commission on Environment and Development (WCED, 1987). Not only is there a strong moral argument in favour of adopting farming systems that are more environmentally sustainable, but there are also moral imperatives to use cropping systems and agricultural knowledge more fairly, focusing on the supply of knowledge, patented systems and appropriate agricultural methods to those in need, especially in less economically developed countries (LEDCs). These issues are more fully explored in the Nuffield Council's reports on the social and ethical issues associated with GM crops (Nuffield Council on Bioethics, 1999; 2003). In those reports, the Council concludes that there is a moral imperative to use all available technologies in the search for more sustainable ways of feeding the poor and the hungry.

How does the advent of transgenic technology impact on the morals and ethics of plant and animal breeding?

Since the late 1980s transgenic technology has made it possible to transfer genes not only from one phylum to another but across the natural barriers that exist between the plant, animal, bacterial and fungal kingdoms. Transgenic techniques are now routinely used in research laboratories worldwide, mainly for investigating gene function and for the production of transgenic microorganisms to be used in industrial processes and in medicine. It is only relatively recently that these tools have been applied to crops and livestock, initially developed by commercial companies to produce crops for use in industrialized farming in more economically developed countries (MEDCs), and fish in aquaculture. Research and development are now moving from the private sector to public institutions, especially in LEDCs.

Moving genes across species barriers is not new; it happens naturally in hybridization. Major crops such as maize and wheat were developed thousands of years ago by interspecific hybridization. But moving genes across phyla and kingdoms is seen by many as 'unnatural' and morally wrong. For example, there is some controversy surrounding moving genes from animals into food plants, partly because strict vegetarians may have objections on moral or religious grounds, but also because some people believe that the process goes beyond normal moral limits (Cooley, 2004). Even in the United States, where biotechnology in agriculture is highly developed and widely accepted, over half the consumers surveyed by Hoban (1999) objected on moral grounds to the production of transgenic animals. Similar objections have been found in other parts of the world (Blaine *et al.*, 2002). It is difficult to see how the view that transgenesis is unnatural and morally wrong can be defended, because genes are lengths of a chemical that appear to have no intrinsic animal or plant qualities (plants and animals have many genes in common), and as demonstrated earlier, there is little 'naturalness' in what humans have done over the past 5000 years to the genetics of crops, livestock, pets and garden plants. There is also strong evidence that genetic sequences are transferred between kingdoms by viruses, and that this has played an important part in evolution. Even the proteins that allow plant cells to replicate appear to have originated in viruses (Villarreal and DeFilippis, 2000). The evolution of mitochondria, the powerhouses of plant and animal cells, involved the wholesale transfer of bacterial genomes, allowing advanced life to exist in an atmosphere growing increasingly rich in oxygen from plants (Gray *et al.*, 2001).

Without a wholesale assault on naturalness, humans would never have achieved the global dominance they currently enjoy. That is not to say that we should not have a greater understanding of the adverse impacts of what we have done to our ecosystems; we should use our existing knowledge of nature (and acquire more) in order to feed and clothe our rising population without further damaging biodiversity, ecosystem functionality and water and soil resources. Ideally we should take this principle further by using our knowledge to repair ecosystem functionality as far as possible. There are probably far too many humans on Earth to be able to live in complete harmony with nature, but we should make a determined and serious attempt to greatly lessen our impacts, using all human knowledge and technologies, including the safe use of transgenic technology.

The debates about the use of GM organisms in agriculture have often centred on the technology itself, questioning not only the safety of the techniques used to construct GMOs, but also on the morality of 'interfering with nature' to this extent and in this way (see Heinberg, 2001). This latter issue has emerged as a controversial area in the debate especially in Europe. Clearly humans have been interfering with nature for many millennia, and as Wolpert and Goldsmith (2000) and others have argued, scientific knowledge (including technology) can be seen as morally neutral until used either for good or for evil. Although others such as Joseph Rotblat (1995) and Edward Goldsmith (see Wolpert and Goldsmith, 2000) disagree with this view, it is difficult to argue that a technology *per se* can be either morally right or wrong. The key questions should be how and by whom knowledge is exploited, what products are made using it, and how and where those products are deployed. Rotblat (1995) has made a strong case for scientists taking moral responsibility for how scientific knowledge should be used, although Lewis Wolpert (Wolpert and Goldsmith, 2000) argues that civil society should, and does, decide how technology is eventually used. This is as true of transgenic technology as it is of nuclear technologies that can be used either to make awesome weapons, or to cure cancers. There will always be serious issues about the morality of producing and using specific transgenic products, and about their ethical use. These issues will be emerging over the next century not just in agricultural biotechnology but also in the use of biotechnology in medicine and industry. To help resolve these issues it is essential that scientists and technologists engage in ongoing debate with civil society about the morality of developing specific transgenic products, which in the future could include 'designer organisms' quite unlike anything we have seen before.

Whatever the moral rectitude of using transgenic techniques in agriculture, there remains the question of whether there is a need for ethical approaches to research and development in this area. Much public and political disquiet about using transgenic technology to produce GM crops revolves around the question of risk. For the crops currently on offer, public attitudes may be summarized as large companies taking the profits, but consumers and the environment being asked to take the risks. Not surprisingly consumers and environmentalists question whether they should be the risk takers. There are of course risks associated with all forms of crop breeding, and it is not unusual for crops produced in conventional breeding programmes to be rejected because they are unstable, perform poorly or are more toxic than accepted varieties, as was the case with high erucic acid oilseed rape varieties now only used for industrial purposes (Ward *et al.*, 1985). For these reasons crop breeding institutes have their own screening processes in place and independent regulatory systems such as the UK National Listing procedures closely scrutinize any crop before it is placed on the market (e.g. Defra, 2006b). It is very rare for unsuitable crops to get through this scrutiny - they usually come to light when a very small proportion of the population unexpectedly proves to be allergic to some component of the crop or food derived from it. However, this regulatory regime has traditionally only dealt with direct impacts of the crop itself, with no consideration of the 'indirect' environmental, social and economic impacts of cultivation systems enabled by new crop varieties. In Europe, for example, conventionally bred novel crops (and new agrochemicals) have enabled a switch from spring to autumn sowing, resulting in substantial damage to the breeding birds and wild plants that rely on farmland for their habitat (Defra, 2006c) and increased soil erosion risk in sensitive contexts (see Boardman, this volume).

There are also ethical issues arising from concentrating the power of new and expensive breeding technologies into the hands of corporations and institutes, leading to a potential loss of research and development of 'public good' crops and animals (Ahloowalia, 1999; AEBC, 2005). Likewise, issues are raised by the potential loss of traditional breeding skills by farmers, coupled with the erosion of genetic diversity in crops and livestock (Jayaraman, 2006). These aspects are especially acute in LEDCs, such as India, where patenting parts of crop genomes has been a controversial subject (FAO, 2003; Jayaraman, 2006). The main ethical issue is whether a wide choice of seeds and other plant and animal material should continue to be freely available to farmers. The perception is that seeds and animals for breeding are part of a farmer's right to access natural materials. As we have seen earlier in this chapter, such materials are arguably not part of the natural world but are derived from it, so there is an argument that farmer breeders who develop such germplasm should have the right to use it, whether or not a commercial company may have a patent on the material. There is a worldwide agreement, the International Treaty on Plant Genetic Resources for Food and Agriculture (FAO, 2004), that goes some way towards setting ethical standards in this area by defining farmers' rights.

Ethical codes for the development of novel crops and cropping systems have not developed in the same way as those for other areas of biological sciences. Until very recently there was a 'laissez faire' attitude to research and development in agricultural science, leading to an 'anything goes' culture in both commercial and public sectors, with few ethical codes to map boundaries and set targets for sustainability of cropping systems and biosafety (Johnson, 2003). One such code in the area of GM crops was developed by the UK Advisory Committee on Releases to the Environment (ACRE), which developed *Guidance on Principles of Best Practice in the Design of Genetically Modified Plants* (2001), which sets out some targets for practices to minimize risks at the earliest stages in crop development. Other than the codes relating to animal welfare, this is a pioneering attempt by a regulatory body to set ethical boundaries for research and development in agriculture more generally, potentially increasing the likelihood of public acceptability and regulatory success for those who conform to the standards set by ACRE. Conforming to ethical codes can be crucial in regulation of medicine, enhancing prospects of technology transfer by reassuring both customers and the political system. Given the rejection of GM technology by consumers in Europe, it might be timely to develop ethical codes for the development and use of all novel technologies (including transgenic technologies) in agriculture. These could encompass issues such as best practice in the production of safe crops and food, animal welfare, sustainability of management systems, and knowledge distribution and access.

Conclusions

Moral and ethical issues in agriculture are not new but have been more sharply focused by the advent of transgenic technologies, partly because novel organisms can be produced more quickly but also as they have characteristics and risks that are unfamiliar to consumers and regulators. There is a consensus across most societies and cultures that humans have the moral right to conduct agriculture but not at the expense of future resources: sustainability is emerging as a new morality. Some sustainable and productive agricultural systems exist in small plot farming and wetland rice production and are seen by many as morally superior to industrialized agriculture, but these are the exceptions; much of our food is produced in unsustainable systems. We urgently need smarter and more sustainable agricultural methods, not more of the same kind of agriculture we have practised over the past century.

The moral case for preventing the use of transgenic techniques in agriculture is weak and somewhat overwhelmed by a pressing need to achieve greater sustainability, given an expanding world population and the damage being done to biodiversity, water resources and soils by 'conventional' agriculture. There are moral issues associated not so much with the technology itself, but with use of transgenic products. Each product, whether a crop, a tree, a fish or a farm animal, raises its own moral and ethical issues, many of which may be specific to a particular culture and place.

Agricultural research and development needs to be conducted within ethical frameworks that set scientific, environmental and socio-economic boundaries and targets for what is and what is not acceptable for society and regulatory systems at any given point in time. If developed in partnerships between scientific communities and civil society, and regularly reviewed, such frameworks could enhance biosafety and public acceptability, leading to more rapid and focused technology transfer. There would be less wasted investment, and researchers and funding agencies would have more confidence in the long-term goals and products of research programmes.

Acknowledgements

I thank my former colleagues Alastair Rutherford of Natural England and Anna Hope of the UK Joint Nature Conservation Committee for their valuable and constructive comments on an early draft of this chapter.

References

Advisory Committee on Releases to the Environment (ACRE) (2001) *Guidance on the Principles of Best Practice in the Design of Genetically Modified Plants.* Guidance Note 13, Defra, London, UK.

AEBC (2005) *What Shapes the Research Agenda in Agricultural Biotechnology?* DTI Publications, London. UK. [Accessed Dec 2006] Available from http://www.aebc.gov.uk/aebc/reports/research_agendas_final_report.pdf

Allard, R.W. (1999) *Principles of Plant Breeding*, John Wiley & Sons, New York, USA.

Amman, K. (2000) Director of Bern Botanical Garden, Switzerland. Personal communication.

Animal Welfare Act 2006 (England and Wales) The Stationery Office, London, UK. [Accessed Dec 2006] Available from http://www.opsi.gov.uk/acts/acts2006/ukpga_20060045_en.pdf

Ahloowalia, B.S. (1999) Trading in Genes and Genetic Resources for Plant Breeding, *New Agriculturalist on line.*[Accessed Dec 2006] Available from http://www.new-agri.co.uk/99-2/perspect.html

Blaine, K., Kamaldeen, S., and Powell, D. (2002) Public perceptions of biotechnology. *Journal of Food Science* 67(9) 3200-3208.

Body, R. (1982) *Agriculture, the Triumph and the Shame.* Temple Smith, London, UK.

Brannigan, M.C. (2001) (ed.) *Ethical Issues in Human Cloning - Cross-Disciplinary Perspectives.* Seven Bridges Press, New York, USA.

Carson, R. (1962) *Silent Spring.* Penguin Books, London, UK.

Conner, J.R, Hamilton, W.T., Sheehy, D.P., Stuth, J.W. and Kreuter, U.P. (1998) Grassland-based livestock production in temperate zones. *World Animal Review* 90(1) 6-13. [Accessed Dec 2006] Available from http://www.fao.org/DOCREP/W8600T/w8600t03.htm#TopOfPage

Conway, G.R. and Pretty, J.N. (1991) *Unwelcome Harvest.* Earthscan, London, UK.

Cooley, D.R. (2004) Are Transgenic Organisms Unnatural? *Ethics and the Environment* 9(1) 46-55.

Daily, G.C. and Ehrlich, P.R. (1992) Population, sustainability, and Earth's carrying capacity. *BioScience* 42, 761-771.

Darwin, C. (1859) *On the Origin of Species by Means of Natural Selection.* Chapter 1 - Variation Under Domestication. John Murray, London, UK.

Defra (2005) *Compendium of UK Organic Standards.* Defra, London, UK.

Defra (2006a) *Farm Animal Welfare: on-farm.* [Accessed Dec 2006] Available from http://www.defra.gov.uk/animalh/welfare/farmed/on-farm.htm

Defra (2006b) *Farming: Crop Varieties.* [Accessed Dec 2006] Available from http://www.defra.gov.uk/farm/crops/varieties/index.htm

Defra (2006c) *Wild Bird Populations 2005: Sustainable Development Strategy and Public Service Agreement Indicators.* [Accessed Dec 2006] Available from http://www.defra.gov.uk/news/2006/061019a.htm

FAO (2003) India sets new rules against biopiracy. *Plant Breeding News* Edition 135. [Accessed Dec 2006] Available from http://www.fao.org/ag/Agp/agpc/doc/services/pbn/pbn-135.htm

FAO (2004) *The International Treaty on Plant Genetic Resources for Food and Agriculture.* [Accessed Dec 2006] Available from http://www.fao.org/ag/cgrfa/itpgr.htm

FAO (2005) *Drought and Salinity Tolerance and GM Crops.* [Accessed Dec 2006] Available from http://www.fao.org/DOCREP/FIELD/006/AD690E/AD690E05.htm

Gray, M.W., Burger, G. and Lang, B.F. (2001) The origin and early evolution of mitochondria. *Genome Biology* 2(6) 1018.1-1018.5.

Heidegger, M. (1977) *The Question Concerning Technology.* Harper and Row, New York, USA.

Heinberg, R. (2001) *Cloning The Buddha: The Moral Impact Of Biotechnology.* Quest Books, Wheaton, Illinois, USA.

Hoban, T.J. (1999) Consumer acceptance of biotechnology in the United States and Japan. *Food Technology* 53(5) 50-53.

Jackson, J.C. (2006) *Ethics in Medicine.* Blackwell, Oxford, UK.

Jayaraman, K.S. (2006) Barriers to building a bigger breadbasket in India. *Science and Development Network* 24 May. [Accessed Dec 2006] Available from http://www.scidev.net/Features/index.cfm?fuseaction=readFeatures&itemid=525&language=1

Johnson, B.R. (2003) Biotechnology: driven by profit or searching for a better environment? In: Serageldin, I. and Persley, G.J. (eds) *Biotechnology & Sustainable Development: Voices of the South and North.* CAB International, Wallingford, UK, 229-237.

Kristensen, T., Soegaard, K., Kristensen, I.S. (2005) Management of grasslands in intensive dairy livestock farming. *Livestock Production Science* 96, 61-73.

Lehmann, V. (1997) The position of Europe's biotechnology industry on bioethics. *Biotechnology and Development Monitor* 32, 67.

Lowe, P., Clark, J., Seymour, S. and Ward, N. (1997) *Moralizing the Environment: Countryside Change, Farming and Pollution.* UCL Press, London, UK.

Mattingly, D.J., Daniels, C.M., Dore, J.N., Edwards, D., Hawthorne, J. (2003) *The Archaeology of Fazzan. Vol.1: Synthesis.* Publications of the Society for Libyan Studies, The Institute of Archaeology, London, UK.

Meadows, D.H., Meadows, D.I., Randers, J. and Behrens, W.W. (1972) *The Limits to Growth: A Report for the Club of Rome's Project on the Predicament of Mankind.* Universe Books, New York, USA.

NIREHG (National Information Resource on Ethics and Human Genetics) (2007) National Human Genome Research Institute, Georgetown University, Washington DC, USA. [Accessed Jan 2007] Available from http://www.georgetown.edu/research/nrcbl/nirehg/

Nuffield Council on Bioethics (1999) *Genetically Modified Crops: The Ethical and Social Issues.* Nuffield Council on Bioethics, London, UK.

Nuffield Council on Bioethics (2003) *The Use of Genetically Modified Crops in Developing Countries.* Nuffield Council on Bioethics, London, UK.

Rotblat, J. (1995) *Nobel Speech to the 1995 Pugwash Conference*. [Accessed Dec 2006] Available from http://www.pugwash.org/award/Rotblatnobel.htm
RSPB (2004) *Assessment of the Sustainability of Industrial Fisheries Producing Fish Meal and Fish Oil*. A report by Poseidon Resource Management Ltd and the University of Newcastle-Upon-Tyne. Royal Society for the Protection of Birds, Sandy, Bedfordshire, UK.
Shoard, M. (1980) *The Theft of the Countryside*. Temple Smith, London, UK.
UNEP (1992) *Rio Declaration on Environment and Development*. [Accessed Dec 2006] Available from http://www.unep.org/Documents.multilingual/Default.asp?DocumentID=78&ArticleID=1163
van Bueren, E.T.L., Hulscher, M., Jongerden, J., Haring, M., Hoogendoorn, J., van Mansvelt, J.D. and Ruivenkamp, G.T.P. (1998) *Sustainable Organic Plant Breeding. Discussion Paper: Defining a Vision and Assessing Breeding Methods*. Louis Bolk Institute, Driebergen, The Netherlands. [Accessed Dec 2006] Available from http://www.ifgene.org/breed1.htm
Villarreal, L.P. and DeFilippis, V.R. (2000) A hypothesis for DNA viruses as the origin of eukaryotic replication proteins. *Journal of Virology* 74(15) 7079-7084.
Walters, L. and Palmer, J.G. (1996) *The Ethics of Human Gene Therapy*. Oxford University Press, New York, USA.
Ward, J.T., Basford, W.D., Hawkins, J.H. and Halliday, J.M. (1985) *Oilseed Rape*. Farming Press Ltd, Ipswich, UK.
WCED (1987) *Our Common Future*. Oxford University Press, Oxford, UK.
Wolpert, L. and Goldsmith, E. (2000) The debate: is science neutral? *The Ecologist* 30(3) 20-23.

Part IV

Practising Systemic Thinking

Chapter 14

Multifunctionality in practice: research and application within a farm business

C. Stoate

Introduction

Whereas previously, demands on agriculture were for arable and livestock products, now they are considerably more varied. Since the late 1990s declining farm incomes have forced farmers to reduce input costs and to explore other sources of income, including diversification of activities on and off the farm. Most recently, increasing commodity prices have provided incentives for arable farmers to increase production. Under these circumstances, greater justification is needed to maintain areas of land for purposes other than cropping. Animal welfare and environmental concerns have also risen to the top of the policy agenda, influencing how farmers manage their land. How farmers react to changing circumstances varies considerably according to their own constraints, opportunities, resources and interests. However, the combination of different objectives is encouraging many to adopt a more integrated approach to farmland management, incorporating economic, environmental and social perspectives.

To some extent, cropping and livestock production have made space for more general natural resource management on farmland, often encouraged by environmental grants such as those available under England's Environmental Stewardship schemes. Legislation, such as cross compliance and the requirements of the EU Water Framework Directive, also contribute to a more integrated approach to resource management on farmland, while Defra's ecosystems approach *Action Plan*, launched in 2007, embodies similar integrated thinking (Defra, 2007). Such integration can lead to multifunctional benefits of single management practices and can require collaboration between farmers, and between farmers and other members of the rural community. Where management is state funded, for example under an Environmental Stewardship scheme, multifunctional benefits can provide better value for the tax payer. Collaboration between actors can contribute to improved social cohesion. However, not all outcomes are positive and there is a considerable need for greater development of this sort of approach. This chapter explores these issues through a series of case studies drawn from the Allerton Project's work in Leicestershire.

The Allerton Project, founded in 1992, has developed through this period of policy change, adapted to it, and informed it through its research and demonstration work. The Allerton Project is based at Loddington (Leicestershire), a 333 ha mixed arable and livestock farm, formerly the home of Lord and Lady Allerton at whose bequest the project was started. Like many land owners, the Allertons combined an interest in rural pursuits such as horse riding and pheasant shooting with farming, in itself a cultural as well as an economic activity. With its interest in shooting and game management, and its long history of research into farmland and woodland ecology, the Game Conservancy Trust (now the Game and Wildlife Conservation Trust) was the obvious choice as the organization to set up and manage the Allerton Project.

The project incorporates its own farm business with a wide ranging environmental research programme, and demonstration activities attracting up to 2000 agricultural professionals to the farm each year. The farm is on clay soils and produces wheat, oats, oilseed rape and beans. Permanent pasture is grazed by a flock of 280 mule ewes. Though some is considerably older, much of the woodland on the farm was planted by the Allertons in the 1960s. This chapter describes the Allerton Project's approach to multifunctionality, based on its own research and in the context of its own farm business.

Examples of multifunctionality

Pheasant shooting

Game management for shooting is a traditional practice on farmland, with grey partridge (*Perdix perdix*) and pheasant (*Phasianus colchicus*) being the quarry species in many areas until the 1960s when numbers of the former started to decline in response to intensification of farming methods and simplification of farmland landscapes. Since then, the focus has been increasingly on release of artificially reared pheasants. However, the game management system applied to grey partridge in the past can also be applied, with some modification, to wild pheasants. This approach was adopted at Loddington. The first year of the project, 1992, was a baseline year, and game management was introduced from 1993. The game management system included the creation and management of many of the habitats researched and developed by the Game Conservancy Trust, supplementary feeding with grain during the winter (September to April), and the control of nest predators such as foxes (*Vulpes vulpes*), rats (*Rattus norvegicus*), crows (*Corvus corone*) and magpies (*Pica pica*) during the nesting season.

Wild pheasant numbers increased substantially from 1993, enabling four or five shoots to be held each year (Boatman and Brockless, 1998). These became an important social event within the local community, with local people being recruited on shoot days as beaters and dog handlers for picking up shot birds. Field trial competitions for dog handlers were also held. Other game species to benefit from the management included red-legged partridge (*Alectoris rufa*) and brown hare (*Lepus capensis*). Many non-game species also increased in abundance. Songbird numbers doubled in the first 5 years of the management, with most of that change taking place in the first 3 years (Stoate, 2002). The birds to benefit most were those species which had declined most nationally over the past 3 decades and were targeted by the UK government for conservation action. In fact abundance of songbirds in spring was correlated with abundance of wild pheasants in autumn and with the number of pheasants shot.

After 2001, the predator control component of the game management ceased in order to assess the contribution of this element of the management system to game and songbird conservation. The gamekeeper transferred to another project and other keepering activities (habitat management and winter feeding) were continued as far as possible by remaining staff. In practice, winter feeding was not carried out as thoroughly as it had been with a full time keeper. The keeper had time, knowledge and commitment that were lacking in staff with other interests and commitments. So availability of winter food as well as predator abundance changed from 2002. Breeding success and abundance of wild pheasants declined substantially when the full management system was abandoned, and shooting was phased out. Songbird numbers also declined slightly, but with some exceptions, not to the same extent. Birds to show a decline were predator-prone species such as spotted flycatcher (*Muscicapa striata*) and song thrush (*Turdus philomelos*).

Not all nest predators increased when the keeper left. Ten pairs of magpies were present in 1992, but numbers were only restored to five, four, five and six pairs respectively in the 4 years after the change. A questionnaire survey of local farmers revealed that nearly half of them had increased magpie control since 1992 (Stoate and Szczur, 2005). In some cases this was because of the work carried out at Loddington and the belief amongst local people that magpie control had been largely instrumental in the increases in abundance of gamebirds and other valued species such as song thrush.

Farm woods as resource

Woodland thinning has been carried out from the start of the project with the dual purpose of improving the habitat for wildlife, and for wild pheasants for shooting. As the woods are primarily plantations that were planted in the 1950s and 1960s, thinning is necessary to create a more diverse structure within the woods, especially permitting the development of herb and shrub layers. However, thinning was stopped in the late 1990s because, as on other farms across the country, farm profits declined from 1996 and the cost of habitat management became difficult to justify.

An opportunity to resume thinning arose when a company (Rural Energy Ltd) was set up in a nearby village, specializing in the installation of woodfuel heating systems and local sourcing of wood chip to fuel them. A fully automated 40 Kw woodchip-fired burner was installed in the Allerton Project's main building in 2004. The building uses about 16 t of wood chip per year, equivalent to an area of about 0.3 ha of thinned woodland. This provides a use for the thinnings produced during woodland management, attaching a value, linked to the price of oil, for a former waste product. Although the burner is a highly sophisticated technology, the chipper is basic equipment that can be run off a tractor pto shaft, making production of the fuel a relatively easy task for farm staff. In practice, as farm staff have also diversified into other activities, some wood chip has been bought in from other local woods, creating a market and encouraging habitat management elsewhere.

In terms of both economic and environmental benefits, sourcing fuel locally is critical. Importing wood chip from further away would increase costs considerably and would increase diesel consumption and therefore carbon emissions. A major advantage of locally sourced wood fuel is that it is almost carbon neutral, contributing to targets for reducing carbon emissions and addressing the issue of climate change.

Woodland thinning benefits a wide range of wildlife including birds such as willow warbler (*Phylloscopus trochilus*) and blackcap (*Sylvia atricapilla*), and plants such as primrose (*Primula vulgaris*) and wood anemone (*Anemone nemorosa*), and rarer species such as herb paris (*Paris quadrifolia*). These are vulnerable to grazing by muntjac deer (*Muntiacus reevesi*) which have colonized the area since the 1970s. A recent survey of eight large SSSI woods revealed that 10 - 24% of bluebell plants were damaged by muntjac at the start of the flowering season (C. Stoate and A. Berry, unpublished data). Although not currently present in sufficient numbers to cause substantial damage at Loddington, densities in some parts of southern England are sufficiently high to result in major ecological damage. Muntjac also browse regenerating coppice stools and therefore reduce the capacity for timber production in coppiced woods.

Wildlife habitat management

Many of the habitats designed for gamebirds and developed and implemented at Loddington are now funded more widely under agri-environment schemes as a result of this research. Wild Bird Seed Mixtures are a development from game crops designed to provide food and cover for gamebirds through the winter. When appropriate crop species are grown they simultaneously provide the same resources for a wide range of farmland bird species (Stoate *et al.*, 2004). Conservation Headlands (selectively sprayed outer edges of commercial crops; Sotherton, 1991) were designed to provide low densities of arable weeds, supporting insect food for gamebird chicks. However, they also provide important insect food for many other farmland bird species during the breeding season when high protein food is essential to nestling survival. At Loddington, conservation headlands and other habitats incorporated into the cropping system enable cereals to be grown as ‘conservation grade’ for which a premium is obtained on sale of the crop.

Perennial grass margins provide a nesting habitat for gamebirds, but also for other birds such as whitethroat and yellowhammer (Stoate *et al.*, 2001). Because they form a dense sward of permanent vegetation, they also help to reduce the occurrence of competitive annual weeds of arable crops such as cleavers (*Galium aparine*) and sterile brome (*Bromus sterilis*) (Boatman *et al.*, 1994). Grass margins have been lost from most farms because they are perceived as a source of annual weeds, but it is clear that they can be used to control such weeds, while simultaneously benefiting wildlife. The dense mat of dead vegetation that forms at the end of the summer provides a wintering habitat for beneficial predatory invertebrates such as carabid and staphylinid beetles, which help to control aphids in crops in the subsequent spring (Collins *et al.*, 2003). Grass field margins can also reduce pesticide drift into hedge bases, and where placed at the base of a slope, can reduce soil and nutrient transport from arable land to water courses.

Beetle banks (low mid-field banks sown with course perennial grasses) were developed as a means of encouraging beneficial invertebrate predators into the centres of fields to reduce aphid thresholds there (Collins *et al.*, 2003). They have been shown to be effective in this role, and also to provide nesting habitat for grey partridges and harvest mice (*Micromys minutus*) and a foraging habitat for other species (Bence *et al.*, 2003).

Cultivation - minimum tillage versus plough

The period of declining farm profits prompted a search for ways of reducing input costs, including those associated with crop establishment. Minimum tillage has been adopted at Loddington since 2002 in order to achieve this. Minimum tillage involves disturbing the soil on the surface rather than ploughing deep and inverting it. Fewer passes with the cultivation equipment are needed, reducing the costs involved. However, capital costs of equipment are relatively high and the farm at Loddington pooled equipment and labour with a neighbouring farm, spreading the fixed costs over more than twice the area of Loddington alone.

Although the switch from ploughing to minimum tillage was for mainly economic reasons, there are also environmental benefits arising from the change (SOWAP, 2007). Research projects have been investigating this issue at Loddington, and at other sites across Europe, since 2004. Initial general results suggest that earthworm density is higher under minimum tillage than plough and that microbial biomass (especially soil fungi) is also higher under minimum tillage. Some bird species show greater use of minimum tillage than ploughed fields in winter because of seed food availability on the soil surface, and in the breeding season, skylarks (*Alauda arvensis*) in particular, have a longer breeding season and are present in higher numbers on minimum tillage (Field *et al.*, 2007).

Higher soil organic matter, coupled with higher earthworm abundance and a healthy soil fungal community increases the water holding capacity of the soil, retaining soil moisture from crops and reducing runoff from arable land to streams. There are therefore implications of soil management for water quality. Losses of phosphorus and nitrogen in runoff are lower under minimum tillage than plough. These implications for water quality are important as new standards under the EU Water Framework Directive call for improved chemical and ecological status of watercourses by 2015. Diffuse pollution from arable land, especially the loss of phosphorus, is currently a major concern to which minimum tillage may offer a partial solution.

Water protection - buffer strip pools

Grass field boundary strips help to reduce movement of soil and nutrients to water by surface runoff, but fail where the land is contoured or steep, or where field drains carry these materials under the buffer strips into water courses. Phosphorus concentrations in field drain water at Loddington have exceeded the level known to cause ecological problems in freshwater on between 60% and 100% of sampling occasions (Stoate *et al.*, 2006). One solution to this is to create wide grass buffer strips at the base of arable slopes and break field drains into a series of shallow pools within them. Water, together with sediment and nutrients, is therefore held in the pools, rather than flowing directly into the adjacent stream. Phosphorus concentrations in the pools at Loddington are approximately half those of the field drain that feeds them (Stoate *et al.*, 2006).

As well as reducing sediment and nutrients in watercourses, the pools have a conservation value in their own right. Although they are inevitably eutrophic, carrying higher concentrations of phosphorus than good quality wetland habitats, many species can survive in them. Between 20 and 60 species of aquatic invertebrate have been recorded in the pools at Loddington, including six nationally scarce water beetle species (Stoate *et al.*, 2006). The rank vegetation around the pools also provides a breeding habitat for pheasants, whitethroats (*Sylvia communis*) and reed buntings (*Emberiza schoeniclus*). In winter, the wetland habitat supports snipe (*Gallinago gallinago*), mallard (*Anas platyrhynchos*) and teal (*Anas crecca*), providing opportunities for shooting.

Wider implications

These examples from Loddington show that there is compatibility between environmental objectives when a multifunctional approach is taken. Environmental objectives can also be compatible with economic and social objectives. In fact, one of the messages to come out of the examples above is that all three objectives can be closely integrated. Natural resource management requires knowledge, and the necessary skills to apply that knowledge. Examples from above include the management of regenerating coppice stools, managing muntjac populations, and feeding gamebirds in winter. Such 'indigenous knowledge' is often under-valued with the result that it is lost. Rising house prices in rural areas exclude those on the lowest incomes who often carry out such work.

The example of magpie control illustrates the integration of environmental management beyond the farm boundary. Control of magpies at Loddington, and the increase in gamebirds and locally valued songbird species, stimulated other local farmers to do the same, based on the perception that magpie control contributed to the increases in these species. This reduction of the local magpie population is the likely explanation for the failure of magpie numbers to be fully restored when their control stopped at Loddington. In terms of improving water quality in water courses, this can only be approached at the catchment scale as what any land manager does inevitably affects what happens down stream. For this reason, some recent research and demonstration projects have focused on involvement of local land managers at the catchment scale, an approach also being taken at Loddington for the Eye Brook, a tributary of the River Welland (Stoate, 2005). Exploring the dual interests of cultural and natural heritage with local people within the Eye Brook catchment is intended to enhance a sense of identity and ownership, and therefore long term commitment to the sustainable management of the rural environment. This work involves local history groups and volunteer naturalists, as well as researchers and land managers.

Collaboration between local farmers is clearly essential to address water quality issues, but the same also applies to meeting economic objectives. The collaboration between Loddington and the neighbouring farmer in pooling equipment and labour to enable minimum tillage to be practised provides an example in which crop establishment costs were reduced, thereby increasing gross margins. Moderate performance of the sheep flock at Loddington because of parasite burdens encouraged a similar collaboration. The neighbouring farmer now sometimes grazes his beef cattle on pasture at Loddington, while Loddington sheep are grazed on the neighbouring farm. Both cattle and sheep benefit from grazing fresh ground in rotation and the differences in sward structure associated with the different grazing species benefit wildlife. Sourcing wood fuel from local woods provides another example of increased collaboration between neighbouring farmers associated with a multifunctional approach to natural resource management. Other studies have shown how local knowledge, communication and collaboration, the concept of social capital, are associated with the most successful farm businesses (Appleby, 2004) and with wildlife conservation (Pretty and Smith, 2003). Such an approach enhances social cohesion and sense of ownership of natural resources within rural communities, and therefore the sustainability of environmental management.

Much environmental management on farmland is now funded under agri-environment schemes. These comprise habitat options that have been researched and developed to meet specific conservation objectives. However, the examples above illustrate how several of them are multifunctional, having benefits for water quality, arable cropping, and cultural activities as well as wildlife conservation. The merits of agri-environment schemes across Europe have been questioned in terms of their cost and delivery (Kleijn and Sutherland, 2003). Such criticism has failed to appreciate the multifunctional role of environmental management practices, or their compatibility with farmers' cultural values. For example, farmers with an interest in game management are more highly motivated to carry out habitat management in an appropriate way for wildlife conservation, as well as being more likely to adopt such management in the first place (MacDonald and Johnson, 2000).

Compatibility of agri-environment schemes with farmers' cultural values therefore contributes to sustainability of environmental management, and to better value for the tax payer. Environmental management within agri-environment schemes can also be linked to market opportunities for farm products, with price premia being obtainable for produce associated with environmental benefits. Such market compatibility also contributes to the sustainability of the management practised and is at the core of the multifunctional approach through a combination of public and private benefits.

The multifunctional approach does not always work as smoothly as everyone might like. Some question pheasant shooting on ethical grounds, but Tapper (2005) illustrates how the conservation benefits of game management apply on a very large scale across the country and across habitats. The implications for wildlife of curtailing game management could have serious consequences for many species.

While deer in farm woods are often regarded as a resource, providing recreational stalking and venison for sale, this applies much less to muntjac than to other deer species. This is because the carcass size is very small, because the species has a skulking nature, and because it causes minimal damage to commercially important crops. While 80% of farmers in the Loddington area recognized that there were muntjac on their farms, and 50% thought their numbers were increasing, only 17% perceived them to be a problem. Local deer stalkers vary considerably in their activity and therefore in the number of muntjac they shoot, leaving several 'source' areas for colonization. Although the economic and ecological damage that may develop in the Loddington area is not clear, there is a need to engage local people in addressing this issue of natural resource management and use.

Minimum tillage can be associated with increased herbicide use to control grass weeds which benefit from the lack of soil inversion (A. Leake, personal communication). Where this occurs, economic benefits to the farmer may be lost. We do not know to what extent the processes that reduce soil and nutrient movement to watercourses also do the same for herbicides. Waterlogged ground has a tendency to emit nitrous oxide, a contributory gas to climate change (El-Sadek *et al.*, 2002). We need to know to what extent this also applies to buffer strip pools that are designed to reduce the loss of nutrients to water.

However, the current demands on food production and, increasingly, biofuel, mean that we cannot turn the clock back and farm in a way that is more akin to 18th century practice. Nor can we abandon farmland for food production and continue to import the amounts of food that we do today. The consequences for greenhouse gas emissions, and therefore for climate change, of transporting food around the world are serious, and not sustainable environmentally, and therefore socially, even if they are economically sustainable in the short term (Cormack and Metcalfe, 2000; Pretty *et al.*, 2005). Increasing oil prices over the next half century will dictate that local food production must be compatible with our other objectives for the rural landscape. A multifunctional approach to farmland management that integrates food, fibre and fuel production with environmental and social objectives, provides the solution if adequately researched and practically grounded. The research being carried out at Loddington aims to ensure that this approach continues to evolve.

References

Appleby, M. (2004) *Norfolk Arable Land Management Initiative (NALMI) Final Project Report.* Countryside Agency, Cheltenham, UK.

Bence, S.L., Stander, K. and Griffiths, M. (2003) Habitat characteristics of harvest mouse nests on arable farmland. *Agriculture, Ecosystems and Environment* 99, 179-186.

Boatman, N.D., Blake, K.A., Aebischer, N.J. and Sotherton, N.W. (1994) Factors affecting the herbaceous flora of hedgerows on farms and its value as wildlife habitat. In: Watt, T.A. and Buckley, G.P. (eds) *Hedgerow Management and Nature Conservation*. Wye College Press, Wye, UK, 33-46.

Boatman, N.D. and Brockless, M.H. (1998) The Allerton Project: farmland management for partridges (*Perdix perdix, Alectoris rufa*) and pheasants (*Phasianus colchicus*). *Gibier Faune Sauvage* 15, 563-574.

Collins, K.L, Boatman, N.D., Wilcox, A. and Holland, J.M. (2003) A five-year comparison of overwintering polyphagous predator densities within a beetle bank and two conventional hedge banks. *Annals of Applied Biology* 143, 63-71.

Cormack, W. and Metcalfe, P. (2000) The efficiency test. *Organic Farming* 67, 22-23.

Defra (2007) *Securing a Healthy Natural Environment: An Action Plan for Embedding an Ecosystems Approach*. Defra, London, UK.

El-Sadek, A., Feyen, J. and Ragab, R. (2002) Simulation of nitrogen balance of maize field under different drainage strategies using the DRAINMOD-N model. *Irrigation and Drainage* 51, 61-75.

Field, R.H., Kirby, W.B. and Bradbury, R.B. (2007) Conservation tillage encourages early breeding by skylarks *Alauda arvensis*. *Bird Study* 54(1) 137-141.

Kleijn, D. and Sutherland, W.J. (2003) How effective are European agri-environment schemes in conserving and promoting biodiversity? *Journal of Applied Ecology* 40, 947-969.

MacDonald, D.W. and Johnson, P.J. (2000) Farmers and the custody of the countryside: trends in loss and conservation of non-productive habitats 1981-1998. *Biological Conservation* 94, 221-234.

Pretty, J.N., Ball, A.S., Lang, T. and Morison, J.I.L. (2005) Food costs and food miles: An assessment of the full cost of the UK weekly food basket. *Food Policy* 30, 1-20.

Pretty, J. and Smith, D. (2003) Social capital in biodiversity conservation and management. *Conservation Biology* 18, 631-638.

Sotherton, N.W. (1991) Conservation headlands: A practical combination of intensive cereal farming and conservation. In: Firbank, L.G., Carter, N., Darbyshire, J.F. and Potts, G.R. (eds) *The Ecology of Temperate Cereal Fields*. Blackwell, Oxford, UK, 373-397.

SOWAP (2007) *Soil and Surface Water Protection using Conservation Tillage in Northern and Central Europe*. LIFE03 ENV/UK/000617. Technical Final Report.

Stoate, C. (2002) Multifunctional use of a natural resource on farmland: wild pheasant (*Phasianus colchicus*) management and the conservation of farmland passerines. *Biodiversity and Conservation* 11, 561-573.

Stoate, C. (2005) Land use and aquatic ecosystem protection within an English lowland catchment: The Eye Brook. In: Anon. (ed.) *Sharing a Common Vision for our Water Resources. Proceedings of the 6th International Conference of the European Water Resources Association, 7-10 September 2005, Menton, France.* CD-ROM, European Water Resources Association, Athens, Greece.

Stoate, C., Henderson, I.G. and Parish, D.M.B. (2004) Development of an agri-environment scheme option: seed-bearing crops for farmland birds. *Ibis* 146, suppl. 2, 203-209.

Stoate, C., Morris, R.M. and Wilson, J.D. (2001) Cultural ecology of Whitethroat (*Sylvia communis*) habitat management by farmers: field boundary vegetation in lowland England. *Journal of Environmental Management* 62, 329-341.

Stoate, C. and Szczur, J. (2005) Predator control as part of a land management system: impacts on breeding success and abundance of passerines. *Proceedings of the International Union of Game Biologists conference*, Braga, Portugal.

Stoate, C., Whitfield, M., Williams, P. and Driver, K. (2006) Wetland creation and mitigation of water pollution from field drains: use of buffer strip pools within an arable landscape. In: Davies, B. and Thompson, S. (eds) *Water and the Landscape: The Landscape Ecology of Freshwater Ecosystems*. International Association of Landscape Ecology and Pond Conservation, Oxford, UK, 331-334.

Tapper, S.C. (2005) *Nature's Gain*. The Game Conservancy Trust, Fordingbridge, UK.

Chapter 15

Linking environment and farming: integrated systems for sustainable farmland management

C. Drummond and C. Harris

Introduction

From 1st January 2005 there has been a radical reform in agricultural policy across Europe, with priorities shifting from production-based farming to those that deliver environmental outcomes. This is a significant step for better resource management, but as with any change there needs to be a balance between environmental care and economic viability and this has played a critical part in the development of farming systems and specifically the promotion of Integrated Farm Management (IFM) systems. Indeed as we face new priorities and changes for farming over the next 10 years, it is important to ensure that sustainable farming systems are flexible enough to account for society's changing demands but at the same time practical enough to be adopted by farmers. This has been particularly important for the work of LEAF (Linking Environment And Farming), a charity set up in 1991 to develop and deliver sustainable farming systems. There are strong consumer and policy signals over issues such as water quality, environmental care, animal welfare, biodiversity, waste management, climate change and energy use, as well as health and obesity issues. Trying to achieve a logical focus and balance to all these issues in the short and long term is complex. This chapter highlights what sustainable farming systems can deliver and demonstrates how IFM, as one system, has the ability to provide a whole farm, practical solution for the majority of farmers in Europe and across the globe.

Sustainable farming systems - what can they deliver?

The term 'sustainable farming' embraces a wide range of issues and objectives, including the role of farming in rural communities; the need for greater protection of the environment; concerns about rural land use; animal welfare; development of local food markets; and the need for farming to support other sectors of the economy, such as tourism. For the purpose of this chapter sustainable farming is defined as farming that 'meets the needs of the present generation without compromising the ability of future generations to meet their own needs' (WCED, 1987: 144). It is a method of agriculture that attempts to ensure the profitability of farms while preserving the environment. Sustainable agriculture integrates three main goals:

environmental stewardship, farm profitability, and prosperous farming communities, underwritten by a fourth key component, political commitment. These goals have been defined by a variety of disciplines and may be looked at from the vantage point of the farmer or the consumer. IFM aims to bring about beneficial opportunities for all by balancing this range of external and internal pressures facing farmers. Table 15.1 provides a range of opportunities linked to long term sustainable objectives.

With the radical reform of the CAP farmers are embracing the environmental challenges, but as we move towards more sustainable farming systems how do we make the environment pay? Farming systems need to be developed that are delivered through:

- Better efficiency.
- Direct support payments.
- Market value and opportunities.
- Trust and communication.
- Striving for sustainable consumption.

There is an intricate relationship between the environment and farming. It is important to recognize the position of farmers in delivering a range of goods, some of economic value and some not measured or measurable by cost alone, such as the joy of bird song or a beautiful landscape. This brings together the mix of natural, human, social, manufactured and financial capital.

Table 15.1: Long term sustainability objectives

Natural Capital	Protect and enhance biodiversity Protect and enhance natural resources Minimize the use of non-renewable energy sources to help achieve long-term carbon-neutrality Embed life-cycles and systems thinking
Human Capital	Health and safety indicators Training, continuous professional development and personal development Innovation
Social Capital	Involvement in the local community Promote positive benefits of the farming community Achieve demonstrable best practice Aim to eliminate all negative impacts on surrounding communities
Manufactured Capital	Minimize inputs of energy, water and other raw materials Minimize waste in production Maximize process efficiency at all points in the value chain
Financial Capital	Secure current and future performance through returns, investments and high level performance Full cost sustainability accounting deployed through the business and valued by the customer Manage supply chain progressively and proactively

As society becomes more urban and wealthier we are more inclined to eat into our natural resources. While technology may in some cases mitigate the full impact of our actions and in some cases undo the environmental footprint and impact of the changing habits of society, we do need to be much wiser in our use of these resources and understand the true cost and 'value' of our activities. Food is central to all our lives and production systems and consumption patterns have a major impact on the environment.

Efficient consumption of food must thus internalize the cost of production and the environment. This requires looking at production cycles over the long-term, as with farming systems such as IFM. The benefits of such approaches should be clear when true recognition is given for other 'values' beyond economics, including social and environmental capital. Furthermore, marketing and advertising also play an important role in shaping consumers' desires and expectations. Retailers have an important influence at the point of purchase; awareness-raising and educational forces can highlight the links between shopping choices and global benefits. Advertising could be exploited to make consumers aware of the consequences of their choices. Encouraging a real change in behaviour has to be easy, fun, 'painless' and of benefit to people's personal lives. Recognition also needs to be given to current achievements otherwise reaching out for more sustainable consumption patterns may appear too difficult.

Although cheap food is frequently still a priority for purchasing decisions, the market for ethical goods is growing. Food is not just another commodity, it is the foundation of personal wellbeing and is inextricably interwoven into a nation's culture, character and land use. In that regard, farming and food production embody a set of skills and capabilities on which the long-term security of any nation still depends. The landscape created by farming creates the conditions necessary for the success of other sectors, especially tourism, and has important impacts on recreation and enjoyment. Farming also has impacts on our health, through the nutritional quality of food. What happens on farms has major implications for both local and global environments.

The expectations for sustainable farming systems are high. Sustainable agriculture must:

- Produce safe, healthy food and non-food products in response to market demands, now and in the future.
- Enable viable livelihoods to be made from sustainable land management, taking account of payments for public benefits provided.
- Operate within biophysical constraints and conform to other environmental imperatives.
- Provide environmental improvements and other benefits that the public wants - such as re-creation of habitats and access to land.
- Achieve the highest standards of animal health and welfare compatible with society's right of access to food at a fair price.
- Support the vitality of rural economies and the diversity of rural cultures.
- Sustain the resource available for growing food and supplying other public benefits over time, except where alternative land uses are essential in order to meet other needs of society.

With the complexity and ever-changing requirements expected of farming and food production there is no one single farming system that can deliver sustainable production. With the principal objectives set out as above, sustainable farming can be delivered through a range of mechanisms that may require additional incentives, markets or infrastructure to be delivered. Fulfilling the requirements of sustainable agriculture is not very easy and changing the farming system cannot be achieved overnight. Sustainable farming is a long term programme; it has internal solutions to internal problems and is responsive to feedback.

One such approach is through IFM, a cropping and livestock production strategy in which the farmer seeks to conserve and enhance the environment while economically producing safe, wholesome food. Its long term aim is to optimize the needs of consumers, society, the environment and the farmer. IFM is a form of Environmental Management System (EMS), a systematic approach that can be used by any enterprise to improve continuously its business management to achieve efficiencies and better environmental outcomes. Defra notes that:

> On an individual farm basis, Integrated Farm Management (IFM) aims to integrate biological processes into modern farming practices using advanced technology to provide the basis for efficient and profitable production which is economically viable and environmentally responsible (Defra, 2003: 130).

Since the ultimate goal is long-term sustainability, IFM optimizes the efficiency and profitability of crop and livestock production while minimizing risks to human health and to the natural and created environments. IFM requires farmers to make complex decisions based on a sound understanding of the potential for interaction among various economic, social and technical factors on the whole farm. As a sustainable farming system IFM is important because it strives to reduce inputs into production systems while maintaining productivity and quality.

One of the prime objectives of developing IFM was to establish a framework which encompasses aspects of farming that provide consumers with the reassurance of environmental protection, animal welfare and the quality, safety and traceability of their food. It is not about quick fix solutions but about evolutionary change and development, offering a practical, realistic and achievable solution for the majority of farmers. As a whole-farm management system it combines the needs of the farmer, the environment, and society. IFM offers the farmer the chance to identify opportunities and threats and to respond to consumer demands for good value food of high quality, which is healthy, wholesome and grown with concern for the environment. Traceability is a further demand from society, a challenge which IFM can address.

IFM also offers the flexibility required to manage legislative matters and meet governmental targets but not lose sight of the farmers' own targets, needs and expectations of their business. IFM is not based on a set of parameters but on informed management processes. This multifunctional nature of IFM as a farming system includes attention to detail and managing all resources available as well as their interactions between each other. This process identifies any adverse effects of farming such as leaching, soil erosion and damage to habitats and biodiversity and measures which need to be taken to encourage their reduction.

For the farmer the IFM approach offers opportunities at all levels. LEAF, the main guardian of IFM in the UK, has over 70 demonstration farms and innovation centres for the farmers to visit and discuss detailed, practical, technical developments and research about IFM. The demonstration farms are where individuals champion the cause of IFM and act as ambassadors for promoting the industry as a whole to the general public. The innovation centres are where research into IFM and its components is pushing forward new solutions for farmers to put into practice. LEAF's work also includes the development of technical tools to assist farmers in delivering IFM. The LEAF Audit is a self assessment management tool for farmers to assess their risks and opportunities and to set targets and actions (LEAF, 2006). The LEAF Speak Out campaign (LEAF, 2005) has provided training and skills to assist in communication and marketing and more recently the development of the LEAF Marque has provided an opportunity for farmers who have delivered environmental and biodiversity benefits to be recognised through the market place.

It is the attention to detail of IFM that also delivers farm profitability. Research into IFM through the 1990s was brought together within a group known as IACPA - the Integrated Arable Crop Production Alliance. The IACPA studies (1995) showed maintenance of profit through the adoption of IFM mainly due to lower input costs. A 40% reduction of crop protection volume, a 15% reduction of fertilizer volume and a 10% reduction in operating costs resulted in a 2% (-20% to +15%) increase in gross margin above conventional production techniques. The research also showed that profitability improves in the IFM system as grain prices fall. More recent work carried out by Forum for the Future (2005) on a LEAF innovation centre demonstrated the logic of IFM in delivering key sustainability benefits through the adoption of minimum tillage where this is the most appropriate choice for cultivations.

Working with the research sector of IACPA, LEAF has built up opportunities through demonstration farms and by providing proof of farmers' actions and delivery of change through the LEAF Audit. Introduced in 1993 in response to farmers wanting to measure their performance and set targets for a more integrated approach, the experience from the LEAF Audit, an environmental management system and self-assessment management tool, has allowed us to track change and specifically improvements over time with farmers setting targets for action and benchmarking (LEAF, 2002).

The audit has developed substantially since it was introduced in 1993 and it is now an interactive CD-rom, updated every year and with key references and help facilities. Farmers who have completed the LEAF Audit have shown consistent, meaningful improvement and aspire to follow best practice. In 2004 LEAF carried out some work on behalf of the Voluntary Initiative, a group established in 2001 by representatives of farming bodies and the crop protection industry to help reduce the environmental impacts of pesticides (LEAF, 2004). The key finding from the analysis was the thoughtful management by farmers and self-assessment through the LEAF Audit. Analysis of the LEAF Audit statement 'We consider cropping practices which encourage biodiversity' showed a steady increase in the number of farmers responding positively, from 45% in 1994 to 70% in 2002. Evidence indicates that farmers' priorities are consistently shifting to greater care for the environment. This aspiration for best practice is down to the level of awareness of the strengths, opportunities and potential of the farm and the key areas of recognition and improvement which the LEAF Audit helps farmers to identify each year.

Marketing value and opportunity

Further drivers of change for delivering sustainable farming systems are where farmers choose to gain market advantage and differentiation by demonstrating a point of difference, either through specific environmental performance, as with the LEAF Marque or organics, or through promoting regional and locally sourced produce. LEAF Marque farmers take forward the messages of IFM, undertake the LEAF Audit and follow specific guidelines with independent, external verification. LEAF Marque provides farmers with the opportunity to be proud of their farming and environmental performance. Increasingly those involved find this partnership motivating and rewarding, and are constantly looking to develop their businesses into new outlets. Indeed all fresh produce retailed by Waitrose now carries the LEAF Marque. Such markets are likely to become of growing importance. This ranges from markets looking for continuity of supply to those which become involved in the production process, sharing the costs of inputs, providing technical guidance and agreeing acceptable farming practices and the timing of production.

While the bulk of production is likely to continue to be for commodity markets an increasing number of farmers are looking to enhance the return from their resources by serving niche markets. Localization of supply attracts some customers, and for suitably placed farmers the direct sale of food to consumers through farmers' markets or local shops can move it from the commodity to the product markets. Such direct marketing activities create a new set of legal obligations. This again emphasizes the versatile role farmers play and one which has been embraced with enthusiasm to manage industry and consumer expectations.

Food markets are exposed to competition, however some resources may also be used to meet demands for services which cannot be imported. Increasingly an affluent society is prepared to pay for recreation and leisure facilities. Part of this demand may be by organizations who wish land to be used in ways which meet their preferences. Thus conservation agencies may pay farmers to provide habitat, or even, in terms of the numbers of some bird or mammal they can show they have harboured. Others may pay for the right to organize war games or to use land for moto-cross, the recreation of historical battles or to hold raves or rallies. Educational markets may also become of increased importance as researchers seek facilities for their activities and schools seek to provide children with first-hand evidence relating to biology and environmental sciences. The traditional markets may also grow; bed and breakfast, fishing, shooting and various forms of hunting are all markets which may generate revenue. These essentially 'club' goods depend upon restricting use or access to those who have paid and a balance will have to be struck between seeking funds from public sources and targeting such sectional markets.

Evidence from the Co-op Customer Survey (2004) suggests that consumers are increasingly willing to pay for environmental goods (see Table 15.2). This is a positive trend but it is also important to make customers recognize the impact of their spending and to continue to explain the true cost of cheap food in terms of food safety, employment, transport distances, animal welfare and wildlife.

Likewise the Institute of Grocery Distribution (IGD) (2005) found that 87% of the population consider farming to be important to Britain and see it as a significant part of British heritage. However, the challenge is to transfer this patriotic view of the countryside into the purchasing of British food. This level of growing commitment is a key part of sustainable consumption, whereby consumers increasingly understand the impact of the decisions that they make on others. With LEAF Marque there is the opportunity of completing a sustainable system, connecting the consumer with the food they are buying, and the story behind its production. This brings alive the excitement and commitment of the farmer, demonstrating a healthy vibrant farming system that supports the soil, crops, livestock, local community and the environment.

Table 15.2: Consumers' attitudes towards ethics of the food industry, 1994 and 2004

	Proportion of respondents	
	1994	2004
General concern about ethical issues - are you more concerned now than in the past?	57%	64%
Willingness to pay more - are you willing to pay a little extra for ethical alternatives?	62%	84%
Wildlife welfare - very important to support products not harmful to wildlife	59%	70%

Source: Co-op Customer Survey (2004)

The LEAF Marque continues to develop with stakeholders, including four of the main retailers (Waitrose, Sainsbury's, Morrisons and Marks and Spencer) as well as farmers, environmentalists and industry groups. Though to date only Waitrose and a newer retailer 'Fresh and Wild' are using the LEAF Marque logo in store, more and more farmers are getting accredited and are using the logo in farm shops and on their products. There are now over 60 product lines and over 150,000 ha registered as LEAF Marqued. The LEAF Marque gives consumers the choice to buy affordable food produced by farmers who are committed to improving the environment for the benefit of wildlife and the countryside. In addition, the LEAF Marque offers full traceability on the food label so consumers can visit 'the farmer' virtually on the web (www.leafmarque.com), access a buyers and suppliers 'matchmaking' service and have the opportunity to visit the farms. Independent joint inspections are being carried out in partnership with other farm assurance schemes, in particular the Red Tractor Assured Food Standard, which was set up to demonstrate food safety standards. Independent joint inspections save money and give credit to those farmers who want to go further in demonstrating best environmental practice with the LEAF Marque.

Conclusion

We could not be more fortunate than within the food and farming industries when it comes to getting involved with consumers. We have a beautiful countryside, local food heroes, celebrity chefs and farmer profiles in magazines all of which conjure up images in our minds, some of them associated with holidays and special occasions, some of them of a past countryside. However consumers and their buying power have changed dramatically over the past 20 years. Globalization, the retailers, choice and availability all mean that the production of food and the story of farming are distant from the food we eat.

Bridging the gap between farmers and consumers is critical to LEAF in order to build understanding and trust. In the long run it is important to get farmers to address consumer concerns and encourage consumers to realize the practicalities and impact of their desires and wishes. LEAF endeavours to understand the concerns people have as consumers and to help them translate these into actions in supermarkets as customers. Experience of non-farming visitors going onto LEAF demonstration farms indicates their main consumer concerns as wildlife and landscape followed by food quality and safety, and animal welfare.

LEAF is looking to develop these farms further to help create the link between purchasing decisions and the impact on the countryside, and develop and promote patterns of consumption that reduce environmental stress and encourage greater value. This is a logical extension to work on sustainable farming. We intend to build on our existing knowledge, experience and communication channels, to demonstrate the importance of making this connection. In particular we will look to support the British farmer and increase public understanding of their important role in the countryside.

Defining the farming systems needed to deliver sustainable agriculture is important. This chapter has aimed to demonstrate the multifaceted requirements of a sustainable farming system and has particularly focused on the potential of IFM and the work of LEAF to deliver this. IFM is not the only system that delivers sustainable farming, but it is one that can bring together the decision-making processes required on farms today in order to deliver the requirements of the market place, consumer, and environment. The suite of activities, tools and experiences which LEAF has put together has meant that farmers can deliver more sustainable practices and communicate this to the public.

References

Co-op Customer Survey (2004) *Consumers' Attitudes towards Ethics of the Food Industry: 'Shopping with Attitude'*. Co-operative Group, Manchester, UK. [Accessed Jan 2007] Available from http://www.co-operative.co.uk/EasysiteWeb/getresource.axd?AssetID=36346&type=Full&servicetype=Attachment

Defra (2003) *Agriculture in the United Kingdom 2003*. The Stationery Office, London, UK.

Forum for the Future (2005) *Farms for the Future*. [Accessed Jan 2007] Available from http://www.forumforthefuture.org.uk/publications/farmsforthefuture_page216.aspx

Institute of Grocery Distribution (2005) *Connecting Consumers with Farming and Farm Produce*. Commissioned by The Sustainable Farming and Food Strategy Implementation Group. [Accessed Jan 2007] Available from http://www.igd.com/downloads/Curry%20Report.pdf

Integrated Arable Crop Production Alliance (IACPA) (1995) *Integrated Farming: Agricultural Research into Practice*. MAFF, London, UK.

LEAF (2002) *LEAF Audit Results 2001/2*. [Accessed Jan 2007] Available from http://www.leafuk.org/sources/4000134/4001069/4029676/audit.htm

LEAF (2004) *Measuring Change Environmental Improvement on Farm over 10 Years, Demonstrated by Analysis of the Leaf Audit Data*. LEAF for the Crop Protection Association. 4.10.2004. [Accessed July 2007] Available from http://www.leafuk.org/sources/4000134/4001069/4032604/Measuring%20change%20-%20LEAF%20report%20final%20version%202.0.pdf

LEAF (2005) *Speak Out*. [Accessed Sept 2007] Available from http://www.leafuk.org/leaf/organisation/speak.asp

LEAF (2006) *The New LEAF Audit User Guide 2006*. 16.8.2006.[Accessed Jan 2007] Available from http://www.leafuk.org/sources/4000134/4001069/4035217/The_Leaf_Audit_User_Guide.pdf

WCED (1987) *Our Common Future*. Oxford University Press, Oxford, UK.

Chapter 16

Environmental and profitable sustainability in agricultural production

A. Williams

Introduction

Producing food is a major concern and trying to make or keep its production sustainable is critical to all our futures. We also must strive to reduce environmental burdens and impacts. Apart from the intrinsic damage to the environment that may be caused by agriculture, rectifying aspects like water pollution is expensive and global warming has consequences that may be incalculable. Environmental sustainability is a relative concept, with few absolutes. It is generally easier to show what is not sustainable than what is.

This chapter concentrates principally on the environmental aspects of sustainable farmland management with some consideration of economic processes at the farm level. There are two strands to the environmental analysis, based on pollution and resource use. Pollutants from agriculture are burdens on the environment that contribute to impacts, such as eutrophication. Eutrophication from agriculture arises mainly from nitrate, phosphate and ammonia; global warming arises from nitrous oxide and methane. Acidification comes from ammonia, which is washed to earth as a weak base and is oxidized in the soil to nitric acid: a strong acid. The other type of burden on the environment is the consumption of resources such as fossil fuels and minerals, both of which have finite reserves.

The economic pillar is essential, however, in that farming must be profitable to continue. This may be helped by subsidies or market intervention at national or international level. To farmers themselves, the sustainability of profits is a critical requirement, but is influenced by many factors and drivers. The maintenance of profit typically results in farmers responding to external pressures, such as regulation, by adapting the farming method to meet new economic conditions. Apart from the generality of economic forces, international agreements to reduce environmental pollution are translated into specific regulations affecting agriculture. The main ones which will or have caused tighter regulation are the EU Water Framework Directive (2000/60/EC) (affecting nitrate, phosphate, faecal indicators); the 1997 Kyoto Protocol in force from 2005 (affecting greenhouse gas emissions); the United Nations Economic Commission for Europe's Gothenburg Protocol 1999, ratified by the UK in 2005; and the EU National Emissions Ceiling Directive (2001/81/EC) (affecting ammonia emissions) (Defra, 2007).

Regulation is likely to decrease profits as remedial measures often incur costs. Solutions to this may involve farmers changing production methods and introducing new technology with the aim of reducing emissions to the environment. The critical question is whether such regulations actually succeed in reducing the emissions of the target species without exacerbating emissions of other species and while not reducing farm level profitability excessively. Another vital question is what are the burdens of producing agricultural commodities? Can they be reduced and how do they affect sustainability?

Modelling approaches to agricultural sustainability - MEASURES and LCA

This chapter introduces two modelling approaches for analysing aspects of agricultural sustainability. The models are outlined then some results are presented and discussed. The work was started at Silsoe Research Institute, but the research team later moved to Cranfield University.

A whole farm model approach for analysing aspects of economic and environmental sustainability is the MEASURES framework - Multiple Environmental Outcomes from Agricultural Systems (Williams *et al*., 2002). For a given area of land, with a defined soil type, rainfall, the prices of crops, operations, fertilizers and machinery, the model finds the most profitable combination and area of crops to grow in a rotation. MEASURES operates at the whole farm level and identifies interactions between profit and pollution from agricultural operations. It allows alternative technologies to be compared, but it does not quantify the resources used to produce the useful agricultural commodities.

Resource quantification is possible using environmental Life Cycle Assessment (LCA). This has been developed for agricultural applications by Audsley *et al*. (1997) and applied to nine agricultural commodities by Williams and colleagues (2006). LCA calculates not only pollutant emissions on the farm but also those from the provision of all inputs and machinery to the farm.

How MEASURES works and what it does

MEASURES takes the definition of a whole farm, which is described in a set of equations and information in databases, and calculates the most profitable cropping rotation for that farm. It also calculates a set of environmental burdens from that farm (Williams *et al*., 2002). The farm may be arable only or mixed. It is based on the Silsoe Whole Farm Model (Audsley, 1981). A farm is described by area, soil texture, rainfall and animal types and numbers. A set of crops can be selected, each with a set of time-bound cultivation requirements that require inputs of labour, machinery, fertilizers, pesticides, etc. Each crop is defined by a growth equation, linking yield to nitrogen (N) fertilizer rate and soil texture. Yield and/or cost penalties are applied if a crop is established too late or too early and if the same crop is grown in successive years. Various rules determine the level of penalty and some crop successions are prohibited to minimize disease transfer. If grazing livestock are included, part of the farm must provide the forage component of their diet. The model finds the optimum plan by maximizing the whole farm net profit (gross margin minus the labour and machinery costs) which does not include the general overheads, typically £200-350/ha. Within this process it assimilates the costs of all inputs for the possible crops, the timings and time taken for each operation together with yields as affected by non-optimal activities. The optimization is achieved using linear programming.

The long term

MEASURES analyses the farm with a long term view, which requires operating in a steady state manner (Audsley, 1981). This is manifested, for example, by the N balance being maintained over a whole rotation, although variations in soil N status may occur within a rotation. An implication of this is that we oblige all N losses to reach the environment, rather than being locked up in short or medium term soil pools (Audsley *et al.*, 1997). This can cause emissions like nitrate leaching to be a bit larger than is normally expected. The philosophy behind this is that all organic N (e.g. from crop debris or manure) will eventually be mineralized and partitioned between useful crop offtake and wastage by senescence, denitrification or leaching. This method thus accounts for all N flows as if the farming systems were to be continued indefinitely, whereas a shorter term analysis would not fully account for, say the fate of organic N in manures (Sandars *et al.*, 2003). Long term N balances were obtained using the SUNDIAL (Smith *et al.*, 1996) simulation model (Williams *et al.*, 2002). Denitrification to nitrous oxide (N_2O) was based on the IPCC 1997 method as used in the UK Greenhouse Gas Inventory (Baggott *et al.*, 2004).

Soil, rain and crop interactions

Crop yields depend on soil texture and, in the case of grass, on rainfall. Rainfall also affects leaching and denitrification. Soil texture is defined by a numerical index (0.5 to 2.5) representing the range of textures from sandy to heavy clay and yields increase with the soil index, because the water retentiveness increases, so supporting better growth in the summer. Working heavier soils, however, requires more time and fuel than lighter ones (Chamen and Audsley, 1993). The hours that a soil is workable decrease as soil index and rainfall increases. The extra flexibility of lighter soils is thus partly offset by lower yields.

Optimal timing

There is an ideal time to establish a crop that maximizes yield. Yield is reduced by premature establishment (disease) or late establishment (reduced growing season). Establishing all of a crop at the ideal time for maximum yield requires a large input of labour and machinery and so increases the cost. Spreading out the time of establishment reduces both yield and costs, but provides the optimum approach. Winter crops are generally higher yielding than spring counterparts and are grown for preference, but the high intensity of autumn crop establishment can mean that growing some spring varieties may be optimal.

Most operations are time-bound within seasonal windows and some must fit into particular sequences, e.g. drilling must follow cultivation. Additional constraints, like Nitrate Vulnerable Zone (NVZ) regulations, can impose restrictions on the timings or duration of operations, e.g. manure spreading. Since these operations are time-bound, the constraints will impose a stress on the whole farm system. A consequence is likely to be a change in the crop rotation to accommodate the constraint. This is a particular strength of the whole farm approach as apparently counter-intuitive responses can result from it with good reason.

Mixed farms

Animals add an extra dimension, mainly through manure. It is assumed that all non-forage feed is bought into the farm, while all arable products are exported. The animal enterprises thus introduce a profit, the plant nutrients in manure, the effort needed to spread manure and gaseous emissions from the animal house and manure management. There are also losses of N by ammonia volatilization, nitrate leaching and crop yield enhancement to consider. Manure management requires extra operations and a variety of constraints, e.g. limits in the application rates and timing. Grazing animals incur these, as well as the need to calculate a forage-based ration with a consequent need for grass.

The N, P and K cycles

The N cycle is crucial to the agronomic aspect of the model as well as the calculation of environmental burdens. A foundation of the model is that it analyses a farm in steady state, so that the flows of N, Phosphorus (P) and Potassium (K) into the farm (feed and hence manure, atmospheric deposition and mineral fertilization) equal those leaving in crop and animal offtakes and environmental burdens for a whole rotation. The model will balance N and K almost exactly, although P may float. The P balance cannot always be maintained exactly, because of manurial inputs and the supra-responsive application rates that are suggested for potatoes, even under MAFF's fertilizer recommendations, RB209 (MAFF, 2000). This means that an excess of mineral P can still give a crop response, but the harvest will not remove it all, so a P surplus can be created.

The overall balance

For a whole rotation on an all-arable farm, the losses thus derive mainly from the difference between the achieved yield, hence N offtake, and total N inputs, which highlights the importance of a high N utilization efficiency.

Annual costs

All costs, whether machinery, labour or chemical inputs, and crop income are considered as annual values to make them comparable. Machinery costs are annualized using discounted cash flow analysis (Audsley and Wheeler, 1978), which requires values for interest rates, inflation, machinery life, machinery price and resale value and maintenance costs. Outputs from MEASURES were tested by comparison with census data and research observations (Williams *et al.*, 2004). They generally agreed well, supporting the validity of MEASURES for analysing whole farms and their management response to optimize profitability.

Analysis of the impacts of NVZ regulations on a pig and arable farm using MEASURES

Scenarios were created to study the outcomes of policies to reduce ammonia emissions and nitrate leaching, applying the rules for NVZs (Defra, 2002). These used a base case of 250 ha with a 2000 pig-fattening unit. The farm was constrained to follow all the manure management guidelines in the Codes of Good Agricultural Practice (MAFF and Welsh

Office, 1998a; 1998b; 1998c) and the analysis considered three soil types (light, medium and heavy) and rainfall levels (600, 800 and 1000 mm/year).

The effects of applying NVZ regulations were investigated using the maximum entry and succeeding limits of 210 and 170 kg total manure N/ha/year, as well as closed winter seasons for manure application. The valuation of manure N was also varied. This value represents the typical proportion of substitution of synthetic N by manure N. It is more particularly the predicted plant-available N in manure (nominally ammoniacal N, plus 10% of organic N, allowing for losses in storage and land spreading). This was set at 50% by default.

Using MEASURES, at the higher manure application limit, the NVZ rules had no effect on nitrate leaching, net profit or any other indicator. At the 170 kg rate nitrate leaching slightly increased, by 1%, and net profit decreased by £3/ha (see Fig.16.1). The main cause of reduced net profit was that 23% more time was required to spread the same amount of slurry on an extra 22 ha at the lower application rate of 170 kg total manure N/ha/year. Adopting the closed season alone had no effect on nitrate leaching.

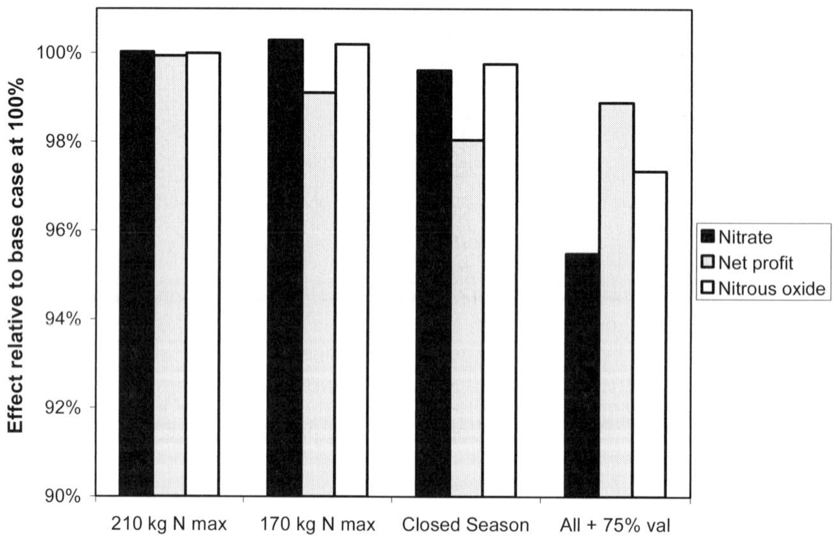

Fig.16.1: Effects of NVZ regulations on nitrate losses, net profit and nitrous oxide emissions on a pig and arable farm

Greater effects were seen when the full, succeeding NVZ regulations were imposed (at 170 kg total manure N/ha/year) together with increasing the fraction of manure N relied upon to 75% (from the default of 50%). This reduced nitrate leaching by 5% and net profit by £3/ha. The same reduction in nitrate leaching could, however, be achieved simply by increasing the manure valuation to 75%, which also boosted net profit by £2/ha. Indeed, increasing the valuation of manure N that can be relied on by farmers has a positive effect on net profit, nitrate leaching and nitrous oxide emissions (see Fig.16.2). In the worst case, if farmers ascribe no value to manure N, nitrate leaching and nitrous oxide emissions are increased by 8% and 5% respectively, while net profit falls by £3/ha. This represents poor, but not unknown, practice. Stretching the value of manure N to 100% reduces nitrate leaching and nitrous oxide emissions by 9% and 5% respectively, while elevating net profit by £4/ha. There were no effects of changing manure valuation on any other pollutants.

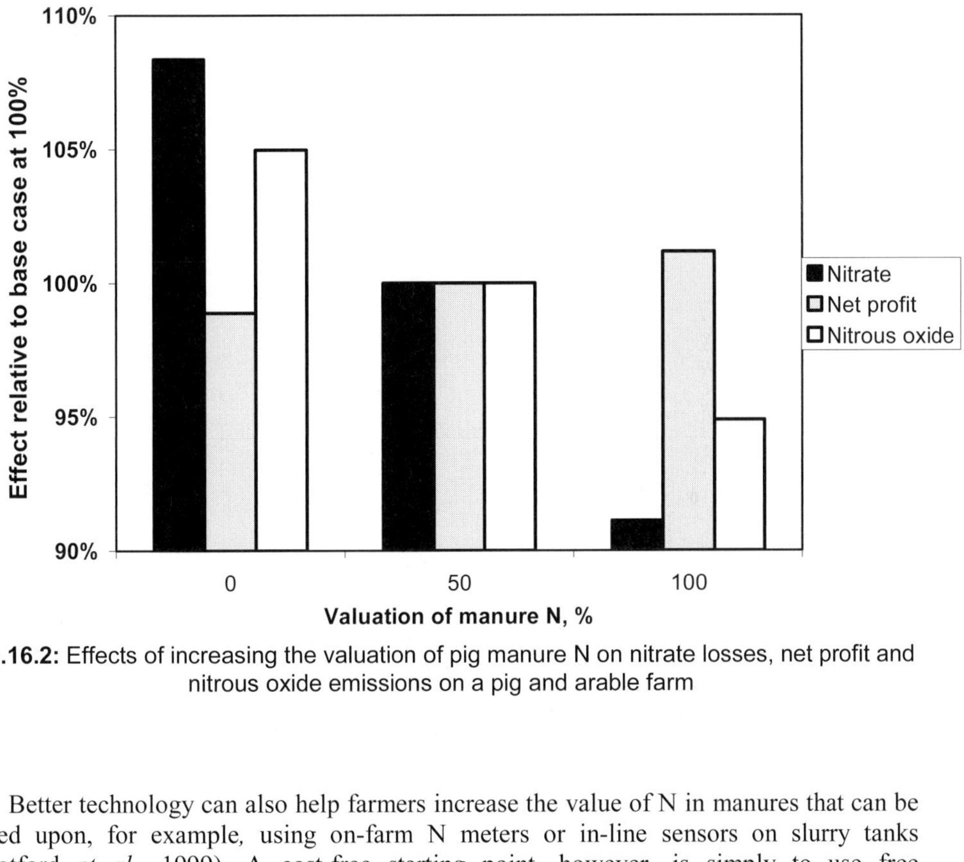

Fig.16.2: Effects of increasing the valuation of pig manure N on nitrate losses, net profit and nitrous oxide emissions on a pig and arable farm

Better technology can also help farmers increase the value of N in manures that can be relied upon, for example, using on-farm N meters or in-line sensors on slurry tanks (Scotford *et al.*, 1999). A cost-free starting point, however, is simply to use free publications, such as the MAFF-sponsored booklets (Pain *et al.*, 1998a; 1998b; 1998c) or a decision support system such as MANNER (Chambers *et al.*, 1999) or SUNDIAL (Smith *et al.*, 1996).

Overall this analysis showed that increasing the value that the farmer can confidently ascribe to the N in applied manure is the most effective way to reduce both nitrate leaching and nitrous oxide emissions and it increases net profit - a 'win-win' solution. This naturally contributes to the economic sustainability of the farm. The imposition of NVZ rules on farms that already follow Good Agricultural Practice can actually increase nitrate leaching and reduce profitability because of the changed cropping that arises from the constraints of NVZ rules. As such, they have a negative effect on environmental and economic sustainability.

Effects of soil and rainfall on abatement costs

NVZ rules are principally intended to reduce nitrate pollution, but the costs of achieving abatement vary with circumstances. We examined the costs of abating nitrate N for the three soil types and rainfalls. The costs of abating nitrate N using NVZ rules were at best cost-neutral (light soil under 600 mm rain) and at worst £4/kg NO_3-N abated (pigs on heavy soil under 600 mm rain). The trends were not clearly systematic over all combinations of rain and soil (see Fig.16.3). This variation resulted mainly from the imposition of a closed season on lighter soils. This caused the cropping to change between soil-rain combinations so that leaching changed. Thus, the analysis incorporates not just the effects of rain and soil on a crop in isolation but also whole-farm responses.

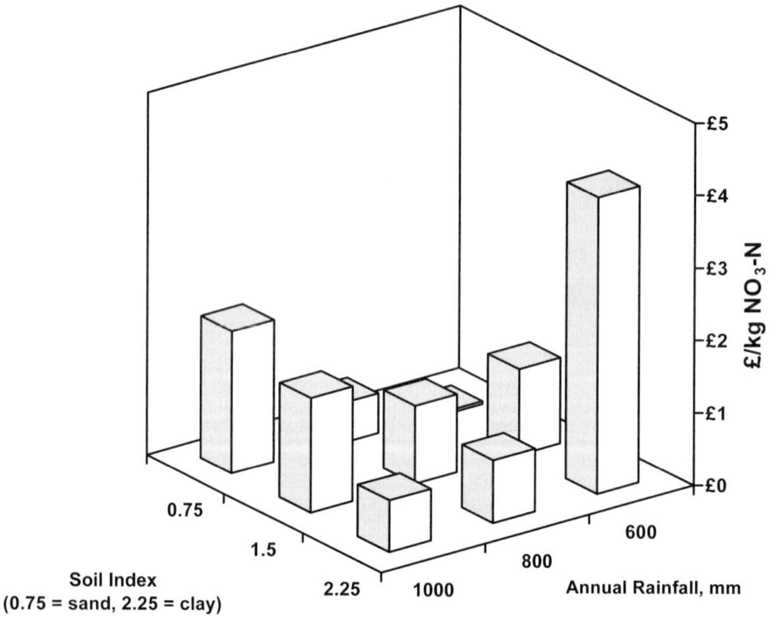

Fig.16.3: Effects of soil and rainfall on the costs of abating nitrate on a pig and arable farm using NVZ regulations

Resource use and environmental burdens in commodity production using LCA

Introduction to LCA

LCA systematically analyses production systems to account for all inputs and outputs that cross a specified system boundary (Fig.16.4). The useful output is termed the functional unit, which must be of a defined quantity and quality, for example 1t bread-making wheat. It is produced using natural resources (e.g. coal, minerals) that may enter the system boundary directly, but more commonly go through a manufacturing or processing stage first. There may also be co-products or waste products like straw or used packaging, together with emissions to the environment, e.g. nitrate to water and methane to air. The system boundary in this work was the farm gate with soil to a depth of 0.3 m.

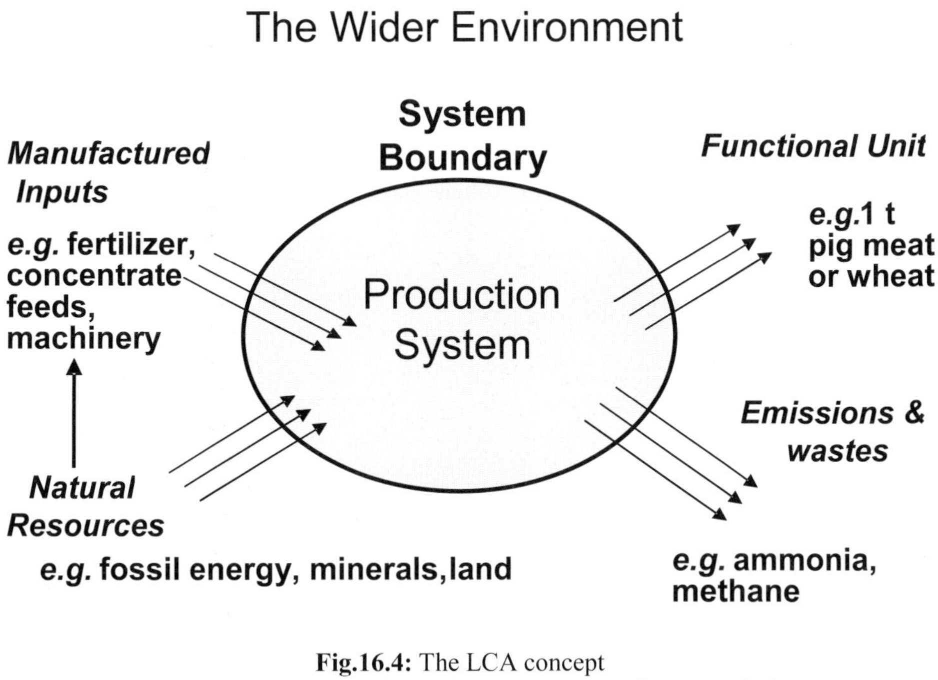

Fig.16.4: The LCA concept
Note that mass flows in and out of the system boundary must balance

All inputs are traced back to primary resources, e.g. electricity is generated from primary fuels like coal, oil and uranium, and the effort needed to extract these is included so that all energy consumption is related to these primary sources. N fertilizers use natural gas as a feedstock and source of energy. P and K fertilizers require energy for extraction from the ground, processing, packing and delivery. Tractors require steel and rubber for their manufacture (embedded energy), all of which incur energy costs, as well as their direct use of field diesel. The burdens incurred in commodity production within a system boundary can generally be described, in the example of energy, as:

$$Burden = \frac{Total\ Energy\ Used}{Total\ Amount\ of\ Functional\ Unit}$$

Analogous expressions apply to other resource usage, emissions or potential impacts. So, while the analysis includes factors like nitrate losses per ha, it involves much more than an assessment of the agricultural emissions on an areal basis. Animals also consume crops and occupy land for housing, ranging and/or nutrition. Their feed crops may be grown elsewhere (e.g. USA, Brazil) and these have their own boundaries. Manure may leave an animal system boundary, but re-enter another crop one. These modules were combined to calculate the burdens of commodity production. Complete details are given in Williams *et al.* (2006) but a summary is presented here.

Constructing a LCA

Flowsheets were produced to describe the main inputs and outputs for each commodity. Many features were generic, e.g. field operations for crops. Analysis of the flowsheets led to the definition of data needs and the relationships needed to connect the data. Some data came from existing Life Cycle Inventories (LCIs) (e.g. Audsley *et al.*, 1997) in which the burdens of producing, for example, agricultural machinery have been evaluated. Others were developed from first principles, with support from the literature and other modelling tools. The sets of data and relationships were assembled in Microsoft Excel spreadsheets.

Functional relationships and the long term

As with MEASURES, N, P and K inputs and outputs must balance in long term rotations and long term N balances were again derived with SUNDIAL. Crop responses to N supply and soil type were modelled using conventional linear exponential equations from long term data from Rothamsted Research's Broadbalk plots. This also included the relationship between crude protein in wheat and N supply, which is important for bread-making quality.

Crop yields and N losses were calculated from a combination of three soil textures and three rainfall levels (derived from Williams *et al.*, 2004) so that a weighted mean of yields and losses could be established at national level (England and Wales). N application rates were derived from the *British Surveys of Fertiliser Practice* (Defra, 2001-2005). Allowances were estimated for the use of manure, in that an apparent under-supply of N is evident from synthetic fertilizer use alone, especially for crops like forage maize. Pesticide use was taken from the Pesticide Usage Survey (Garthwaite *et al.*, 2005), which contains tables of historic use. These were used in quantifying the frequency of applications (field work), the overall usage as dose-ha and the energy of manufacture. The post-harvest activities of crop drying, cooling and storage were also included.

Land use for crop production is reported assuming average yields for Grade 3a land (Bibby and Mackney, 1969). Yields were scaled up or down using linear coefficients derived from Moxey *et al.* (1995) for other land grades. Grass production was modelled using a method based on site class (Brockman, 1995), coupled with clover use (and N losses) from the NCYCLE model of Scholefield *et al.* (1991). This calculated grass productivity (dry matter, metabolizable energy and protein offtake) for both grazed and conserved situations. It included grass from intensive lowland pastures to hills. The field operations needed for forage conservation and grass ley establishment were evaluated by analogy with arable crop production.

Animal production

Arable crops can be considered in relative isolation, but this is not true for animal production. In the simplest cases of eggs, and pig and poultry meat production there are typically breeding nuclei, from which secondary herds or flocks are derived, and these feed replacement genetic material into the commercial sector. Within the commercial sectors, several housing and rearing systems co-exist, each with its own characteristics. Changing the proportions of one part can have several interacting effects on other production areas. The situation with ruminant production (sheep meat and beef) is yet more complex. These may be reared in geographically diverse areas and with a complex network of genetic flow. Beef animals are also partly derived from the dairy sector. In addition, ruminants interact closely with the grassland that supports them.

Animal production was modelled by creating structural system models of the sectors using sets of linear equations to connect the input-output relationships of each part. These included the pyramidal structure of the poultry flocks and pig herds, together with the geographical diversity of sheep and the interdependencies of milk and beef (Sandars *et al.*, 2006). The system models included the needs of the animals for nutrition, housing, grazing or ranging land and direct energy for stock management and milking.

For poultry, pigs and sheep, nutritional requirements were constants for particular sub-systems, but for beef and dairy a model was used, based on the Agricultural Research Council (1980) metabolizable energy (ME) approach. ME needs were related to yield and animal size. The dry matter intake to supply the energy depended on the feed mixture (e.g. proportions of maize silage and grazed grass). Linear equations were solved to find requirements for forage dry matter and concentrates. Enteric methane emissions were scaled by the forage intake.

Additional direct emissions of ammonia, nitrous oxide and methane were also included using emission factors based on those in the UK Inventory (Baggott *et al.*, 2004). Apart from enteric methane, these arise from manure in housing, fields, storage or land spreading. These and the burdens of managing manure were included in the burdens of animal production. Animals were credited with the fertilizer value of manure. This was calculated by a procedure that includes all losses of N and then estimates the response of a crop (e.g. wheat) to that N supply. This was then equated to the alternative burdens of supplying N fertilizer to obtain the same crop response. This is the technique of *displaced burdens* that is used in LCA and allows crop and animal commodities to be modelled separately.

Feed crops were modelled like all arable crops but included those grown overseas. Transport and processing were included, with economic valuation used to partition the burdens of processing where more than one product was produced, e.g. soya beans give oil and meal; wheat produces flour and wheatfeed. Domestic feed crops included feed wheat, barley, rape and field beans. The aim was to include at least 95% of animal feeds currently used either by direct analysis or by analogy. The land requirement for animal production thus includes a component of land overseas.

Aggregation of burdens

Emissions to the environment, whether from farms, industrial processes or transport, were initially quantified by individual chemical species and were then aggregated into environmentally functional potentials for causing impacts. The major ones were global warming potential GWP_{100}, eutrophication potential (EP) and acidification potential (AP). GWP_{100} was quantified over 100 years in terms of CO_2 equivalents. The main agricultural sources (and scaling factors relating them to CO_2) are nitrous oxide (296) and methane (23). The main agricultural sources of EP are nitrate and phosphate to water and ammonia emissions to air. The main agricultural source of AP is ammonia (with some sulphur dioxide (SO_2) from fossil fuel combustion). Ammonia is weakly alkaline but, when deposited or in the atmosphere, it is oxidized to nitric acid. Non-renewable abiotic resource use (ARU) was aggregated using the CML factors (CML, 2005). Many natural resources are on a common scale that is related to the scarcity of the resources and is quantified in terms of the mass of the element antimony (Sb).

Life Cycle Inventories of agricultural commodities

The burdens of producing arable and animal derived commodities are diverse (Tables 16.1 and 16.2) with large systematic differences between arable and animal as well as distinct differences within each set. While each commodity stands alone caution is needed in comparing commodities as their nutritional, cultural and commercial properties differ. Rape incurs more burdens than wheat but contains more protein and much more energy in oil which can be used as a food or biofuel. Potatoes contain about 80% water (compared to wheat at 14%) and their storage is much more demanding, but they fulfil many roles in catering. The high energy demand for cooling uses over 30% of the primary energy used in potato production (compared to 3-5% for post harvest grain). Potatoes thus use relatively less energy as fertilizers (24%) compared with grains (55%) where fertilizer usage is clearly the dominant energy demand. The different energy usage impacts on the contribution of N_2O and CO_2 to GWP_{100}. N_2O contributes 46% for potatoes and 77% for grains. Early potatoes, however, have no farm storage so their N_2O contributes 70% to GWP_{100}. Overall, N_2O dominates agricultural greenhouse gas emissions.

Table 16.1: Main burdens of production of each field crop commodity (per t fresh weight)

Impacts and resources used	Bread wheat	Oilseed Rape	Potatoes
Primary energy used, GJ	2.4	4.9	1.4
GWP_{100}, t CO_2 equiv.	0.70	1.4	0.20
Eutrophication Potential, kg PO_4 equiv.	3.0	8.1	1.0
Acidification Potential, kg SO_2 equiv.	3.3	9.1	0.84
Pesticides used, dose ha	0.92	0.71	0.37
Abiotic Resource Use, kg Sb equiv	1.5	2.8	0.94
Land use Grade 3a, ha	0.14	0.31	0.027

Table 16.2: Main burdens of production of each animal derived commodity

Impacts & resources used *	Beef	Sheep meat	Pig meat	Poultry	Eggs	Milk
Primary Energy used, GJ	27	25	21	15	14	2.6
GWP_{100}, t CO_2	14	14	5.4	4.0	4.2	1.0
Eutrophication pot'l, kg PO_4	120	210	66	33	40	5.0
Acidification pot'l, kg SO_2	300	500	240	96	140	14
Pesticides used, dose ha	1.2	0.9	1.6	1.5	1.2	0.09
Abiotic resource use, kg Sb equiv.	34	29	38	23	36	3.1
Land use (#) Grade 3a, ha	2.3	0.55	0.65	0.61	0.52	0.12
Land use Grade 5, ha	1.3	2.6	0	0	0	0

(*) per t of carcass, 20,000 eggs (about 1t) or 1m^3 milk
(#) Grade 3a includes land that could be used for arable, 5 supports grazing only

The other obvious comparison is that animal production incurs larger burdens per t than field crops. This is no surprise because animals live off crops (of which grass is a special example) and they concentrate plant material into human foods that contain much greater concentrations of higher quality protein. They also contain fat and micro-nutrients. While the energy, GWP_{100} and abiotic resource use are about an order of magnitude larger than for field crops, the impacts on eutrophication and acidification potential are about two orders larger. Much of this can be attributed to ammonia emissions. Ammonia arises ultimately from undigested protein and points to a need to improve protein utilization in order to reduce ammonia related impacts. It occurs both from grazing and housed stock.

Land use by animals is more complex than that for crops which were all scaled to Grade 3a land. Pigs and poultry live on feed from arable land but ruminants can digest the cellulose in grass (and other plants in swards) and so can live on poorer land, especially that in uplands and hills. This feature of ruminants means that much more agricultural land can be successfully used productively than by cropping alone (especially with sheep). One disadvantage, however, is the emission of enteric methane that is (currently) an inevitable bi-product of cellulose digestion. That, coupled with nitrous oxide emissions from grassland, means that the ratios of GWP to energy use are significantly higher for ruminant than non-ruminant meats.

The production of meat by pigs and poultry incurred smaller burdens than sheep and beef (but note that these are not culturally or nutritionally completely interchangeable). The breeding cycles of poultry are much shorter than those of pigs which, in turn, are shorter than those of ruminants (one gestation per year). This (and different feed needs) has resulted in breeding poultry with very high feed conversion efficiencies and growth rates. The feed needed to support the breeding overheads is also small (Fig.16.5). There is a clear trend in overheads reducing from suckler cow beef production (< 1 calf surviving per year), through sheep (> 1 lamb per year overall), to poultry where one hen can give rise to about 250 offspring. The effect of beef from the dairy sector is large. We considered that dairy calves are born in order to stimulate milk production, thus all breeding overheads were allocated to milk production. The large contribution of cross-bred calves from the dairy sector thus reduces the burdens of beef production.

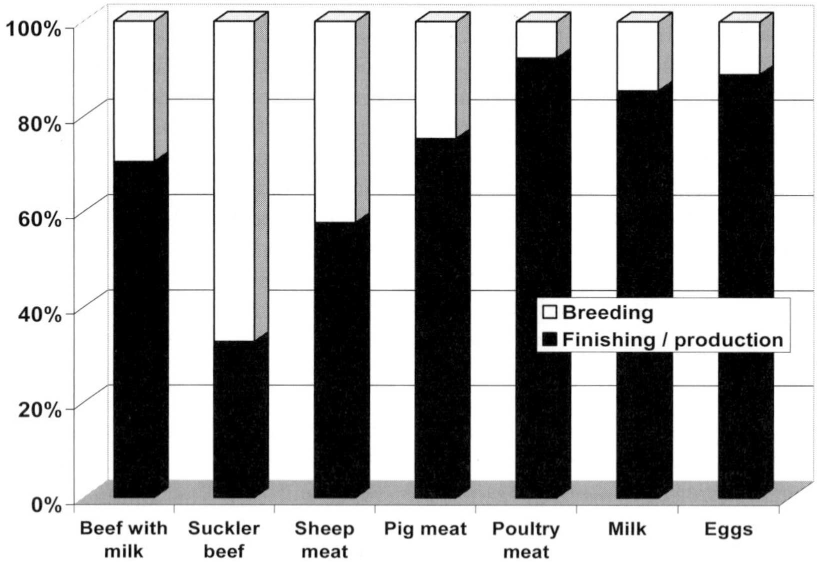

Fig.16.5: Distributions of energy by function in animal production

Discussion

Two methods have been used to analyse the environmental and economic sustainability of agricultural production. The whole farm method of MEASURES showed how policies that were designed to reduce emissions to the environment can backfire. Not only can emissions of other pollutants be increased, but profitability can be depressed so reducing the economic sustainability of farming. A typical consequence is that an established system is squeezed by regulation. In order to recover profits the system is re-optimized. This may, for example, involve growing more spring than autumn sown crops which tend to leach more. This is not an intuitive response to regulation to control nitrate leaching but it is an entirely logical consequence. The analysis also shows how the cost to farms of the same regulation can differ widely depending on rainfall and soil type, which are not in the farmers' control.

Predicting such outcomes is a great benefit of using a whole farm approach rather than relying on the results of a change in one process alone. The effects of farmer responses to changed economic conditions (including regulation) cannot be underestimated and can clearly cause unexpected consequences. They may not always be rapid, however, depending on the need for change to involve capital investment (with low profits also reducing investment rates). The whole farm approach also showed clearly how a changed farmer attitude to the fertilizer value of manure could be beneficial both to the environment and profit. Although not included in MEASURES, such an attitudinal change would reduce resource and primary energy use through reducing demand for N fertilizer. Identifying these doubly beneficial outcomes serves to increase both environmental and economic sustainability. Encouraging the take-up of such practices would seem to be a rational policy goal. The whole farm approach embodied by MEASURES provides a very suitable tool for examining a diversity of novel techniques and policies, whether these policies come from governments or via the purchasing power of large retailers. It is beyond the scope of this work to analyse future trends in profit, but it would seem rational for all major stakeholders to ensure that domestic agriculture remains profitable overall, although some sectors will flourish more than others.

The LCA method is more rigorous than MEASURES and accounts for resource use as well as emissions. One could reasonably infer from the LCA undertaken that nothing is wholly environmentally sustainable *per se*. This follows from the consumption of finite resources, such as primary energy (about 97% non-renewable) and other associated materials that contribute to the unified abiotic resource use scale (e.g. steel for tractors or rock phosphate). It should also be noted, however, that the use of fossil fuel by agriculture is relatively small, being less than 1% of the national total (DTI, 2006). This could lead one to conclude that, if the needs of food production are considered paramount, then the supply of primary energy to agriculture should be protected to permit current production levels to continue for a considerable period. Another supportable argument is for less meat consumption, especially non-ruminant as ruminants can utilize grassland unsuitable for arable production. The environmental burdens are much higher for meat than crops but cultural resistance to any enforced change may be very high.

The analyses also provide a baseline from which to improve the environmental performance of agriculture. Much of the energy going into agriculture is to supply fertilizer which is dominated by N. Improving N utilization in both crops and animals is within the scope of agricultural research and would seem to be an imperative. N production is energy demanding, representing about 50% of the energy used in grain production. Its manufacture still emits the powerful greenhouse gas N_2O (although technology improves to reduce this emission), but field emissions dominate. About 85% of all grain N_2O emissions arise in the field. Indeed N_2O dominates agricultural greenhouse gas emissions (e.g. 77% from grain) so that the sector's contribution to the national total is much higher than the 1% of energy used.

Surplus N also enters watercourses as nitrate and particularly enters the atmosphere as ammonia from livestock. There must still be considerable scope for improvements to be made in nitrogen utilization efficiency. These must be environmentally beneficial and quite possibly will improve profitability (as in the MEASURES example). The manufacturing of N fertilizer itself is beyond my scope but the energy demand for N fertilizer has decreased about five-fold in the last 100 years (Jenssen and Kongshaug, 2003). Further such radical improvements seem unlikely now but the alternative sourcing of natural gas from oil fields where it was previously flared off is a serious possibility. This would reduce the primary energy need for production and the emissions from oil production itself. Whether UK agriculture would wish its N fertilizer supplies to originate from some of the less politically stable oil producing countries of the world is another matter. These two modelling approaches are powerful tools for exploring agricultural sustainability. While environmental sustainability is not infinite it can be extended and it is clearly in our interest to do so.

Acknowledgements

The research on which this chapter is based was funded by Defra and the support of Dr Donal Murphy-Bokern is of particular note. The author is most grateful for this support. The great contribution of Eric Audsley and Daniel Sandars to the research is firmly acknowledged.

References

Agricultural Research Council (1980) *The Nutrient Requirements of Ruminant Livestock.* Commonwealth Agricultural Bureaux, Slough, UK.

Audsley, E. (1981) An arable farm model to evaluate the commercial viability of new machines or techniques. *Journal of Agricultural Engineering Research* 26(2) 135-143.

Audsley, E. and Wheeler, J.A. (1978) The annual cost of machinery using actual cash flows. *Journal of Agricultural Engineering Research* 23, 189-201.

Audsley, E., Alber, S., Clift, R., Cowell, S., Crettaz, P., Gaillard, G., Hausheer, J., Jolliett, O., Kleijn, R., Mortensen, B., Pearce, D., Roger, E., Teulon, H., Weidema, B. and van Zeijts, H. (1997) *Harmonisation of Environmental Life Cycle Assessment for Agriculture.* Final Report, Concerted Action AIR3-CT94-2028. European Commission, DG VI Agriculture, Brussels, Belgium.

Baggott, S., Brown, L.M.R., Murrells, T., Passant, N. and Watterson, J. (2004) *UK Greenhouse Gas Inventory, 1990 to 2002.* Annual Report for submission under the Framework Convention on Climate Change. AEA Technology plc, Culham Science Centre, Abingdon, Oxon., UK.

Bibby, J.S. and Mackney, D. (1969) *Land Use Capability Classification. Soil Survey Technical Monograph No 1.* Soil Survey of England and Wales, Harpenden, UK.

Brockman, J.S. (1995) Grassland. In: Soffe, R.J. (ed.) *The Agricultural Notebook.* 19th edition. Blackwell, Oxford, UK, 194-226.

Chambers, B.J., Lord, E.I., Nicholson, F.A. and Smith, K.A. (1999) Predicting nitrogen availability and losses following application of organic manures to arable land: Manner. *Soil Use and Management* 15(3) 137-143.

Chamen, W.C.T. and Audsley, E. (1993) A study of the comparative economics of conventional and zero traffic systems for arable crops. *Soil and Tillage Research* 25(4) 369-390.

CML (2005) *LCA Resources at CML-IE.* Institute of Environmental Sciences (CML), Leiden University, The Netherlands. [Accessed 2005] Available from http://www.leidenuniv.nlinterfac/cml/ssp/index.html

Defra (2001-5) *The British Survey of Fertiliser Practice. Fertiliser Use on Farm Crops for Crop Years 2000-4*, Defra, London, UK.

Defra (2002) *Guidelines for Farmers in NVZs – England. Revised Edition.* Defra, London, UK.

Defra (2007) Defra website. [Accessed Sept 2007] Available from http://www.defra.gov.uk/

DTI (2006) *Digest of United Kingdom Energy Statistics 2006.* Department of Trade and Industry, London, UK. [Accessed June 2007] Available from http://www.dti.gov.uk/energy/statistics/publications/dukes/page29812.html

Garthwaite, D.G., Thomas, M.R., Dawson, A. and Stoddart, H. (2005) *Pesticide Usage Survey Report 2002: Arable Farm Crops in Great Britain 2004.* Pesticide Usage Survey Group, Central Science Laboratory, Sand Hutton, York, UK.

Jenssen T.K. and Kongshaug, G. (2003) *Energy Consumption and Greenhouse Gas Emissions in Fertiliser Production.* IFS Proceedings Number 509, The International Fertiliser Society, York, UK.

MAFF (2000) *Fertiliser Recommendations for Agricultural and Horticultural Crops. RB209.* Seventh edition. The Stationery Office, London, UK.

MAFF and Welsh Office Agriculture Department (1998a) *Code of Good Agricultural Practice for the Protection of Air.* MAFF Publications, London, UK.

MAFF and Welsh Office Agriculture Department (1998b) *Code of Good Agricultural Practice for the Protection of Soil.* MAFF Publications, London, UK.

MAFF and Welsh Office Agriculture Department (1998c) *Code of Good Agricultural Practice for the Protection of Water.* MAFF Publications, London, UK.

Moxey, A.P., White, B. and O'Callaghan, J.R. (1995) The economic component of NELUP. *Journal of Environmental Planning and Management* 38(1) 21-34.

Pain, B.F., Smith, K.A., Chambers, B., Scotford, I.M., Cumby, T.R. and Nicholson, R.J. (1998a) *Making Better Use of Livestock Manures on Arable Land: Managing Livestock Manures: Booklet 1.* MAFF Rural Marine and Environment Division, London, UK.

Pain, B.F., Smith, K.A., Chambers, B., Scotford, I.M., Cumby, T.R. and Nicholson, R.J. (1998b) *Making Better Use of Livestock Manures on Grassland: Managing Livestock Manures: Booklet 2.* MAFF Rural Marine and Environment Division, London, UK.

Pain, B.F., Smith, K.A., Chambers, B., Scotford, I.M., Cumby, T.R. and Nicholson, R.J. (1998c) *Spreading Systems for Slurries and Solid Manures: Managing Livestock Manures: Booklet 3.* MAFF Rural Marine and Environment Division, London, UK.

Sandars, D.L., Audsley, E., Canete, C., Cumby, T.R., Scotford, I.M. and Williams, A.G. (2003) Environmental benefits of livestock manure management practices and technology by life cycle assessment. *Biosystems Engineering* 84(3) 267-281.

Sandars, D.L., Audsley, E. and Williams, A.G. (2006) A model for calculating the environmental burdens of livestock production. Paper presented at EURO XXI, 21st European Conference on Operational Research, June 2006. Reykjavik, Iceland.

Scholefield, D., Lockyer, D.R., Whitehead, D.C. and Tyson, K.C. (1991) A model to predict transformations and losses of nitrogen in UK pastures grazed by beef-cattle. *Plant and Soil* 132(2) 165-177.

Scotford, I.M., Cumby, T.R., Richards, P.A., Keppel, D. and Lenehan, J.J. (1999) Development of an in-line nutrient sensing system for livestock slurries. *Journal of Agricultural Engineering Research* 74, 303-316.

Smith, J.U., Bradbury, N.J. and Addiscott, T.M. (1996) Sundial: a PC-based system for simulating nitrogen dynamics in arable land. *Agronomy Journal* 88(1) 38-43.

Williams, A.G., Audsley, E. and Sandars, D.L. (2006) *Determining the Environmental Burdens and Resource Use in the Production of Agricultural and Horticultural Commodities. Main Report.* Defra Research Project IS0205. Cranfield University and Defra, Bedford, UK. [Accessed Jan 2007] Available from http://www.silsoe.cranfield.ac.uk and http://www.defra.gov.uk

Williams, A.G., Sandars, D.L. and Audsley, E. (2004) *Environmental Benchmarks of Arable Farming.* Final report to Defra (CSG15) on project ES0112. Silsoe Research Institute, Silsoe, UK.

Williams, A.G., Sandars, D.L., Annetts, J.E., Audsley, E., Goulding, K.W.T., Leech, P. and Day, W. (2002) *Framework to Evaluate Farm Practices to meet Multiple Environmental Objectives.* Final report to Defra (CTE9901) on project WA0801. Silsoe Research Institute, Bedford, UK.

Chapter 17

The use of indicators to assess the sustainability of farms converting to organic production

C. Firth, I. Milla and P. Harris

Introduction

Sustainability in relation to agriculture is increasing in importance as the linkages between economy, society and the environment are more widely recognized (van der Werf and Petit, 2002). The term has a proliferation of definitions and some consider it as so vague a concept that it has little meaning and should be discarded. However, it is commonly applied to ecological or environmental, social and economic aspects of farming systems. Sustainable agriculture is essentially concerned with the ability of agro-ecosystems to remain productive in the longer term. A key feature of farm sustainability is the need to protect and make optimum use of limited natural resources within an economically efficient and socially acceptable agricultural system. Organic farming is often presented as a sustainable solution for agriculture.

In 1999 the Committee on Agriculture of the Food and Agriculture Organization (FAO) adopted a report which concluded that while organic agriculture is only one of a number of approaches to sustainable agriculture and many of its techniques are also practised under other agricultural systems, it is unique in being regulated by laws and certification programmes which prohibit most synthetic inputs and mandate soil-enhancing crop rotations (FAO, 1999). A further advantage of the organic approach noted by Vereijken (1999: 51) 'is that it offers a market model of shared responsibility by producers and consumers for a sustainable and multifunctional management of the rural areas as agro-ecosystems'. The farm system is not only chosen by the farmer (producer) but the customer (consumer) is involved too through asking in the market for products of quality and environmental friendliness and paying a premium for them.

In order to judge whether an agricultural system or farm is sustainable, an easy to use tool or method is required that can provide information understandable to practitioners such as farmers and to other stakeholders such as policymakers. It is a challenge to measure farm level sustainability in a practical way. A number of methods of assessing agricultural or farmland sustainability have been devised, although many of them do not operate at a practical farm level, and are either of a theoretical nature or designed to be applied at a national or district level. This chapter draws on a project by Milla (2003) which set out to discover a practical system of measuring and assessing sustainability and then to test it on a number of farms which had converted to organic production in order to assess whether the change of production system had a positive benefit on sustainability.

Materials and methods

A literature review was conducted in order to examine previous attempts to assess farm level sustainability. Sustainability has no units. In order to assess it and to help monitor progress towards a more sustainable state, an indicator or benchmark is required against which to measure (Bell and Morse, 1999). The wealth of literature on indicators of sustainability, however, includes diffferent interpretations of such indicators and their roles, involving qualitative as well as quantitative measures (Rigby *et al.*, 2001). The definition used by Smyth and Dumanski (1993) of indicators as attributes which measure or reflect the status of the environment or a change in its condition, highlights a strong focus on the environmental aspects of sustainability. Others have attempted to devise more integrated measures or focus on the value of indicators as tools for management (Glenn and Pannell, 1998; Pannell and Glenn, 2000). For example, Pannell and Glenn (2000: 136) argue that 'the value of a sustainability indicator springs from its potential to improve decision making, and so it is best thought of as a source of information'. Following from this they argue that indicators should be targeted to the needs of different actors and purposes. Having undertaken a review of literature, Rigby and colleagues (2000: 8) summarized the desirable properties of indicators in terms of functions and purposes as follows:

- to assess conditions and changes;
- to compare across place and situations;
- to assess conditions and trends in relation to goals and targets;
- to provide early warning information; and
- to anticipate future conditions and trends;
- selection of the most significant information;
- simplification of complex phenomena;
- quantification of information, so that its significance is more readily apparent; and
- communication of information, particularly between data collectors and data users.

They conclude that the development of sustainability indicators is extremely useful in that it pulls the discussion of sustainability away from abstract formulations and encourages explicit consideration of the operational meaning of the term.

Rigby *et al.* (2001) and van der Werf and Petit (2002) both review a range of techniques, with the latter undertaking a thorough review of 12 indicator methods. The methods reviewed varied in that they worked at different scales and levels: field; crop or product; farm; regional or national. They also varied in what they measured, in most cases dealing with environmental dimensions rather than the socio-economic aspects. Finally, they varied in their complexity and time taken to assemble and process the data.

For this study of organic farming it was decided to select a method of assessment which met three key criteria. First, it should include a broad range of the dimensions of sustainability, namely the economic, social and environmental. Second, the method should be able to operate at farm or whole farm system level, which is so important in capturing the integrated nature of organic farm systems. Last, it was important to have a method of assessment which was not too complex or time-consuming.

Thus after examining a number of techniques previously used to measure sustainability it was decided to adapt and test the Method IDEA (Indicators of Sustainability of Agriculture) (Vilain, 2003). This is a quantitative evaluation system devised for assessing sustainability at the farm level in France, which had not been assessed before in the literature. The premise of the approach is that the sustainability of agricultural systems can be evaluated according to their technical, social, spatial and human characteristics. In this system the three main dimensions of sustainability are agro-ecological, socio-territorial and economic (see Table 17.1).

The agro-ecological indicators consider the autonomy of the farm, its use of renewable energy and materials and its effectiveness in minimizing pollution, and they also take account of the natural capital the farm generates. This dimension is sub-divided into three components: diversity (crop and animal); land organization; and agricultural practice; and then into 15 indicators. The socio-territorial dimension involves components of quality of products and territory; employment and services; and ethics and human development; it has 14 indicators.

Sustainable agriculture aims to be socially acceptable and equitable rather than being based simply on crude economic performance. The absence of simple indicators, capable of assessing all the possible complex notions, makes the economic dimension quite complicated. Society is changing in its demands and its exigencies daily. The farm world is evolving and is influenced by society. Economic sustainability is a basic requirement for the sustainable development of the farm. It is a result of the combination of production factors and the interactions between the environment and farm techniques within a market world. The economic dimension of the assessment method used here examines the production system and its development from a financial and economic point of view. This dimension has four components: viability; independence; ownership; and efficiency; and eight indicators.

The three dimensions are separately evaluated to a maximum of 100 units each. Each of these is sub-divided into three or four main components (see Table 17.1) and these are further sub-divided into indicators, which are based on measurable data or values. Each indicator has one or several issues which constitute its value (for an example see Table 17.2). These can never be negative; the minimum will always be zero and the maximum will vary but will be appropriate for each indicator. The boundary levels are thus set on the basis of expertise to represent the point at which significant changes occur. The score +1 represents an elementary unit of agricultural sustainability, in other words a small approximation towards sustainable development. During this project 37 different measures were made to score each farm on a wide range of features: farming practices, land organization, product quality, social interactions on and off the farm, economic performance and efficiency.

Table 17.1: Sustainability dimensions, components and indicators used in the IDEA system

Dimensions	Components	Indicators	Max	Total
Agro-ecological	Diversity	Animal diversity	15	35
		Annual crop diversity	15	
		Perennial crop diversity	15	
		Protected species/heritage	5	
	Land organization	Crop rotation	10	30
		Field area	10	
		Ecological regulation	10	
	Agricultural practice	Fertilization	12	35
		Effluents	4	
		Pesticides	12	
		Animal welfare	3	
		Soil protection	3	
		Soil fertility	3	
		Irrigation	3	
		Energy	3	
Socio-territorial	Quality of the products and the territory	Food quality	8	33
		Heritage and landscape	10	
		Access to the space	5	
		Social interactions	10	
	Employment and services	Direct marketing	5	33
		Multifunctionality	5	
		Contributions to employment	11	
		Collective work	9	
		Forecast farm life	3	
	Ethics and human development	Training	8	34
		Work intensity	7	
		Life quality	8	
		Isolation	8	
		Hygiene, health and safety	8	
Economic	Viability	Economic viability	10	30
		Level of specialization	10	
		Return on capital	10	
	Independence	Financial autonomy	15	25
		Dependence on government aid	10	
	Ownership	Inheritance	20	20
	Efficiency	Productive process efficiency	15	25
		Economic efficiency by inputs	10	

Table 17.2: Example of the process of scoring in IDEA (Farm I)

Annual crop diversity	1996	Score 4	2001	Score 17
Number of species growing	2	4	10	15
If more than six crops	No	0	Yes	2
Legumes in the rotation	No	0	No	0

Table 17.3: Descriptions of the farms

	FARM I	FARM II	FARM III
Location	Warwickshire	Lincolnshire	Bedfordshire
Type of business	Family	Company	Family
Farm size (ha)	36	1956	24
Farm type	Mixed	General cropping	Horticulture
Type of conversion	Single step (1999)	Phased over 7 years	Phased over 6 years
Data based from	1996-2001	1996-2000	1998-2001

Three different farms (see Table 17.3) were chosen from a group which were studied as part of a Defra funded project on 'Conversion to organic vegetable production' (Anon., 2005). This project studied the agronomic and economic performance of a group of ten farms which had converted to organic production with vegetables. The farms chosen for discussion in this chapter represent different farm types from the group: mixed, arable, and intensive vegetable farms. For each of the three farms, Farm Business Survey (Defra, 2003) data had already been assembled as part of the Defra project, for the year prior to conversion then for 4 or 5 years following the intial conversion. We were able to make use of the data from the pre-conversion year and information from 4 to 5 years later as a comparison; this was used to determine the value of the indicators for each of these years. The data were mainly obtained from farm records and accounts and were supplemented by a questionnaire and interview with each farmer, mainly to obtain additional environmental and social information, and by observations undertaken when visiting the farm. This enabled a snapshot of the sustainability of the farming system pre- and post-conversion to organic production. The farmers all engaged constructively in the project and felt the wider view that this project took captured a broader picture of events on the farms during and subsequent to conversion to organic production.

Results

At an individual farm level, changes in the 37 values measured prior to conversion were compared with those measured 4 to 5 years later on the three farms. The changes were mapped in radar diagrams (see Fig. 17.1 for an example) for the individual farms and then combined for overall analysis (Fig. 17.2). Figure 17.1 clearly shows increases in sustainability in all three dimensions for Farm I. Through the process of conversion to organic production and adopting less intensive agricultural systems, two of the farms (see Fig. 17.2) improved their agro-ecological sustainability. This was achieved through stopping the use of artificial fertilizers and pesticides and by growing a wider diversity of crops, all of which was necessitated by the adoption of practices to comply with organic standards (UKROFS, 2001). All the farms adopted a wider range of crops in order to form longer crop rotations which are at the heart of the organic farming system. The rotations of both cash and fertility building crops, such as clover, form the basis for providing nutrients for subsequent cash crops. The rotations also provided opportunities, in the absence of chemicals, for weed, pest and disease control. All of the farms increased their area of grassland, providing improvements to soil protection and fertility. One of the farms (Farm II) was already using less intensive farming methods prior to its conversion and contained a substantial amount of land in set-aside and environmental schemes. This farm was also converting more slowly than the others; therefore its agro-ecological changes were less marked.

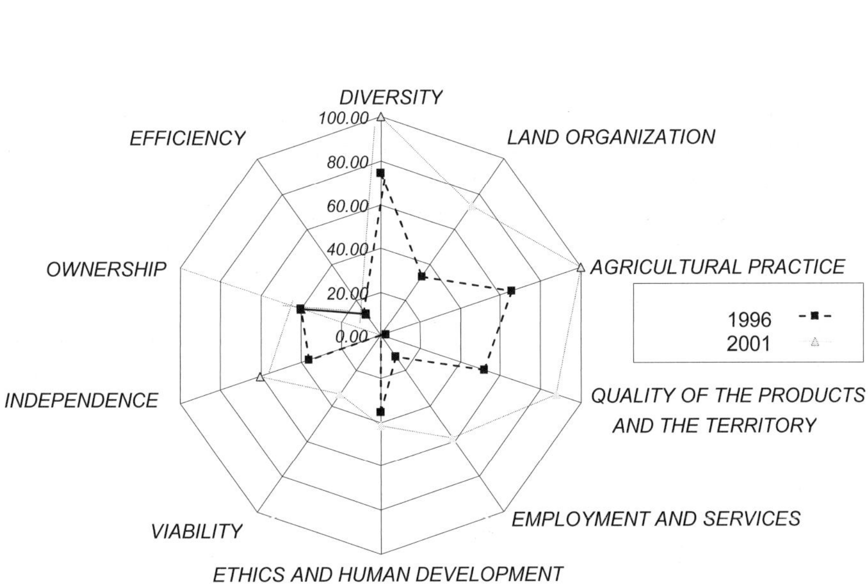

Fig.17.1: Differences in sustainability indicator component values, Farm I 1996-2001

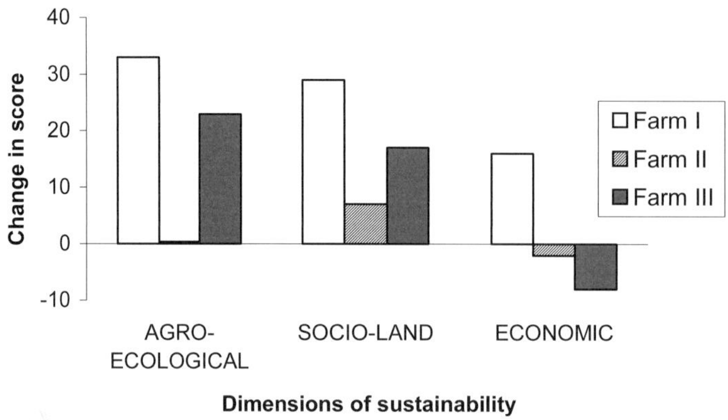

Fig.17.2: Changes in dimensions of sustainability for three farms

All of the farms increased the social dimension of their sustainability through the conversion process. This was related to adopting new marketing practices, such as direct and more local marketing, taken up by farms I and III, with benefits to the local economy, and greater interactions with the consumers. Another contributory factor was an increase in labour employed on the farms, resulting in employment benefits within the local economy. By contrast, only one of the farms increased its economic sustainability. This was the small mixed farm where the premiums to be obtained from the sale of organic products enabled it to become more economically viable. All of the farms incurred costs of converting to organic production through reduced financial output during the conversion period as they took land out of cash crop production and put it into fertility building. The state financed Organic Farming Scheme conversion payments, which the farms received, did not meet all these costs. This is reflected in their reduced economic sustainability. Output levels rose again when the farms started to sell organic products following the in-conversion period. However, after only 3-5 years these did not reach pre-conversion levels. Thus the time period chosen may be too limited to assess the full economic impacts of organic conversion.

Conclusions and discussion

This chapter has found that conversion to organic production has increased the social sustainability of all the case study farms and the environmental dimension of each has tended to either increase or stay at a high level. In contrast the economic sustainability of the farms has either decreased or increased only slightly from a low level. It is thought that the costs of restructuring the businesses and the decline in income during the conversion period have contributed to this decrease. Greater environmental and social sustainability have been achieved by growing a greater diversity of crops, stopping the use of artificial inputs, marketing more produce locally and employing more labour. It is often the case that these benefits to the environment and society do not necessarily bring great rewards to the farmer in terms of higher on-farm economic returns.

The method has provided a useful illustration of changes in sustainability by considering a wide range of factors and proved useful as a tool to engage and discuss the impact of changes in management practices with farmers. The aggregation of the indicators into components and dimensions allowed the results to be displayed in an accessible form. This was achieved in a relatively short period of approximately 2 weeks per farm. It was necessary to make a number of adjustments to the French system of evaluation, which is primarily geared to livestock farms, to adapt it to UK arable/vegetable farms. The study presented here is limited by the small sample of three farms and applying the technique to a larger sample of farms would certainly strengthen the conclusions. In addition, this study considered data over 2 years for each farm and the choice of the years could have affected the economic results as economic performance does vary from year to year according to variations in the weather, prices and so forth. In future it would probably be more advisable to choose average results from a farm in order to make the economic evaluations. Finally, it became apparent during the project that a period of 3 to 5 years from conversion is probably too short a time to assess the full impacts of conversion to organic production on the sustainability of a farm. A longer time scale is therefore recommended.

References

Anon. (2005) *Conversion to Organic Field-scale Vegetable Production*. HDRA, Final report on project OF0191 to Defra, UK.

Bell, S. and Morse, S. (1999) *Sustainability Indicators: Measuring the Immeasurable*. Earthscan, London, UK.

Defra (2003) *Instructions for Collecting the Data and Completing the Farm Return*. Farm Business Survey. Defra, Farm and Animal Health Economics Division, London, UK.

FAO (1999) *Organic Agriculture*. Committee on Agriculture Report 15th Session, January 25-29, 1999, Rome, Italy.

Glenn, N.A. and Pannell, D.J. (1998) The economics and application of sustainabilty indicators in agriculture. Paper presented at the 42nd Annual Conference of the Australian Agricultural and Resource Economics Society, 19-21 Jan, 1998. University of New England, Armidale, New South Wales, Australia.

Milla, I. (2003) *The Use of Indicators to Assess the Sustainability of Farms which have Converted to Organic Production in the UK*. Unpublished MSc thesis, University of Coventry, UK.

Pannell, D.J. and Glenn, N.A. (2000) A framework for the economic evaluation and selection of sustainability indicators in agriculture. *Ecological Economics* 33, 135-49.

Rigby, D., Howlett, D. and Woodhouse, P. (2000) *A Review of Indicators of Agricultural and Rural Livelihood Sustainability*. Sustainability Indicators for Natural Resource Management and Policy Working Paper 1. IDPM, University of Manchester, Manchester, UK.

Rigby, D., Woodhouse, P., Young, T. and Burton, M. (2001) Constructing a farm level indicator of sustainable agricultural practice. *Ecological Economics* 39(3) 463-78.

Smyth, A.J. and Dumanski, J. (1993) FESLM: An international framework for evaluating sustainable land management. A discussion paper. *World Soil Resources Report 73*. FAO, Rome, Italy.

UKROFS (2001) *UKROFS Standards for Organic Food Production*. Defra, London, UK. [Accessed 2006] Available from http://www.defra.gov.uk/farm/organic/legislation-standards/standard.pdf

van der Werf, H.M.G. and Petit J. (2002) Evaluation of the environmental impact of agriculture at the farm level: a comparison and analysis of 12 indicator-based methods. *Agriculture, Ecosystems and Environment* 93(1) 131-45.

Vereijken, P. (1999) Manual for Prototyping Integrated and Ecological Arable Farming Systems (I/EAFS) in Interaction with Pilot Farms. Wageningen, The Netherlands.

Vilain, L. (2003) La Méthode IDEA: Indicateurs de Durabilité des Exploitations Agricoles -Guide d'Utilisation. Educagri éditions, Dijon, France.

Chapter 18

Agri-environment schemes and sustainable farmland management: a farm level assessment of the economic costs and benefits

J. Jones

Introduction

This chapter explores farm level economic considerations that have an important bearing on the financial incentive for farmers to participate in agri-environment schemes. It does this by examining the evidence from three case study farms all of which were all lowland farms with a mix of arable and grazing livestock enterprises (sheep and/or beef cattle) and were participants in both the Countryside Stewardship Scheme (CSS) and the Wildlife Enhancement Scheme (WES). These schemes have now been replaced by the Environmental Stewardship Higher Level Scheme (HLS). The WES was only available on land within Sites of Special Scientific Interest (SSSIs). CSS was potentially available to all farmers but entry was competitive (as with HLS) and only farms with the potential for high environmental benefit were likely to be successful. The farms provide some contrast in farming systems, size, tenure, resources and physical conditions for farming and wildlife.

The original fieldwork was undertaken in 2002 (Jones, 2006). Conditions since 2002 have been altered fundamentally by the reform of the CAP. As a result the calculations have been reworked to account for the change in support payments under the CAP. This provides a clear contrast in the position before and after CAP reform. The decoupling of agricultural subsidies means that cutting back agricultural production to suit an agri-environment prescription is a process that is no longer tied to the cuts in agricultural support payments, which was often the case under the old regimes. It also removes some of the directly associated income benefits or costs that were a feature of the old regimes (such as the leasing of subsidy entitlement quota that became surplus to requirements when stocking levels were reduced). The effects can be subtle and complex but the impact is generally quite profound as the analysis shows and the implications of this are briefly discussed.

Investigation of the relationship between payment rates and income foregone

Few studies in the past have looked directly into the amount of income foregone as a result of applying prescriptions under CSS or WES. There were a number of studies undertaken for the Countryside Commission to establish appropriate rates of payment for the CSS in its early years (Savills, 1994; Produce Studies, 1994). However these are not in the public domain and are now rather dated. More recent studies have tended to concentrate on issues such as participation rates and opinions of farmers. These might be affected by the financial impact of scheme adoption but do not produce any specific evidence as to what the financial cost/benefit might be. The economic evaluation of the CSS undertaken in the late 1990s (Short *et al.*, 2000) only explored payment rates against income foregone by asking farmers whether a reduction in payments would be acceptable. Not surprisingly over three-quarters of them said that lowering of payment rates would be unacceptable (Short *et al.*, 2000: 45). But this could easily have been a behavioural reaction rather than a precise indication of the financial cost/benefit. Other behavioural approaches have all confirmed the prime importance of financial justification (e.g. Wilson and Hart, 2001; Morris and Potter, 1995; Morris *et al.*, 2000). Yet these have not been followed up by farm level analysis of CSS.

There has been more work analysing the impact on farm income of Environmentally Sensitive Area (ESA) scheme participation. The closest parallel has been the work of Tate (2001) in the Shropshire Hills ESA. Thomson and Slee (1998) used a survey-based approach to examine the impact on farm income of ESA participation in Scotland. A broad-based evaluation of agri-environment schemes (CRER and CJC, 2002) looked for statistical evidence to link uptake of prescriptions with the theoretical profit level based on the budgets estimating income foregone submitted to the EU Commission (Defra, 2004a). This proved inconclusive, possibly because the actual income foregone varies so much from the theoretical level.

Notional income foregone and the regulatory framework for setting payment levels

Payment levels provided by agri-environment schemes funded by the CAP European Agricultural Guidance and Guarantee Fund (EAGGF) under the Rural Development Regulation are subject to limits that, at least notionally, restrict the extent to which farmers can profit from them. This applies to CSS but not specifically to WES (although the payment levels under the latter do comply with the same guidelines). Member States must demonstrate compliance with budgets showing the expected level of income loss as a result of following scheme prescriptions. At the time the survey work on the three case study farms was undertaken in 2002 the payment levels were subject to conformity with Article 24 of Regulation 1257/1999 (Council of the EU, 1999). The latest requirements in Article 39 of Regulation 1698/2005 (Council of the EU, 2005) limit payments to a combination of income foregone, additional costs resulting from the commitment given, and 'Where necessary, they may cover also transaction cost'. Previously transaction cost was not mentioned but there was provision to provide an incentive. Regulation 445/2002 Article 19 limited this to 20% of the income foregone and the additional costs (Council of the EU, 2002). If circumstances 'on the ground' matched those in the Department for the

Environment Food and Rural Affairs (Defra) budgets presented to the EU Commission farmers should be able to do little more than cover their costs. The evidence from the three case study farms is that there is often a considerable gap between the cost and income foregone and the payment level allowing farmers to generate income from scheme participation and not just break-even financially.

Methodology

The case study method has very rarely been used to identify income foregone levels associated with agri-environment scheme prescriptions. The few examples tend to produce evidence on a whole farm basis rather than specifically for each prescription within the scheme (Frost *et al.*, 2004). Evidence on costs per prescription was limited to a few isolated cases cited in the farming press (Impey, 2002) and these invariably referred to the single example of the Game Conservancy Trust farm at Loddington.

The method for assessment of the financial consequences of the application of agri-environment prescriptions was to draw up a partial budget for each prescription. Partial budgeting is a conventional approach used widely in farm planning (Turner and Taylor, 1998). The budgets take account of not only income loss and gain but also any cost savings or additional costs incurred. It is the method that was used by Defra in demonstrating the expected income foregone in compliance with Regulation 1257/1999 (Defra, 2004a). There are some quite difficult issues of principle to be addressed in preparing these budgets. These are discussed at some length in Jones (2006). But it is worth noting that decisions were taken to:

- Use actual data based on a single year rather than attempt to forecast, average or normalize.
- Look solely at the net cost of capital works rather than trying to assign any subsidiary benefit to the income or capital value of the farm.
- Include the benefit of the leasing value of any livestock headage payment quota no longer required even though in practice the small scale of such transactions might make them too small to be worthwhile (on their own).
- Ignore any notional (i.e. family) labour costs and any fixed paid labour costs (i.e. not including overtime or casual work) on any labour saved or additional labour required.
- Treat fixed elements of machinery cost such as depreciation, road tax and insurance as irrelevant and only include variable elements such as fuel and repairs.

One important implication of ignoring elements of cost that are fixed and do not alter as a result of a change in management is that family run farms, which tend to have costs that are substantially fixed or notional, differ markedly to farms where contractors are used and costs vary closely in line with production. Although none of the case study farms was run entirely by use of contractors they did make use of contract help to differing degrees.

The effects of decoupling are expected to be far-reaching and might involve significant changes in prices, markets and farm economics generally. There is work indicating the likely impact based on various modelling approaches (summarized in Defra, 2003) with some likely farm level effects reviewed by Jones (2004). However the evidence was not clear enough nor the ramifications simple enough to include this in the adjustment of the partial budget calculations to account for subsidy decoupling. In any event it was thought that it might be clearer to see what effect subsidy decoupling would have with all other conditions held constant - *ceteris paribus*. This did not just involve subsidy removal however. Revenue or costs directly linked to them such as Suckler Cow Premium quota leasing were also removed. But any wider effects, even where they are closely linked to subsidy receipts such as the price of store cattle or agistment fees, were ignored. The Single Payment has been treated as a separate concept not related in any way to decisions made as to what to produce and how to produce it. No specific allowance was made for the effects of applying cross compliance conditions. These costs are believed to be relatively minor in financial terms (Defra, 2004b). The farmers on the case study farms put the compliance cost at almost nil because the conditions overlap so much with existing legal requirements and codes of practice.

Table 18.1: Background information on the case study farms

Characteristics	Farm A	Farm B	Farm C
Size in ha	236	121	828
Tenure	Rented (AHA and FBT)	96% owner occupied	'Landlord–tenant' partnership
Soil type	Sandy to sandy clay loam	Clay loam over shale	Chalk downland
Enterprises	Cereals, potatoes, field vegetables, beef store fattening and sheep on tack	Cereals, beef suckler cows and fattening, breeding sheep and livery horses	Cereals, rape, and herbage seeds, beef suckler cows and fattening and outdoor pigs
Year of entry to agri-environment	1994	2001	1999
Landscape	Riverbank and floodplain	Coastal (cliff tops)	Downland and water meadows
Archaeology	Medieval cattle enclosure	Elizabethan and W.W.II defensive works	Romano-British fort; tumuli; iron age dyke
Notable birds	Barn owl; lapwing; shelduck; curlew	Cirl bunting	Stone curlew; skylark
Other notable wildlife	Otter; red squirrel; water vole; brown hare	Nothing unusual	Butterflies; orchids

The background details of the three case study farms are summarized in Table 18.1. This shows that there was a range of farm size and forms of land tenure. Whilst they were all broadly arable farms with livestock the mix and level of intensity varied. Farm A had intensive arable cropping and the livestock enterprises were small scale and of secondary importance. Farm B had a mix of cereals, sheep, beef and a small equine enterprise. Farm C had significant beef and outdoor pig enterprises as well as a complex arable rotation involving herbage seeds as well as a range of combinable crops. The environmental characteristics of the three farms in Table 18.1 show a reasonable amount of significant wildlife and archaeological interest. They all offered a combination of the characteristics defined in the CSS national target landscapes and across the three farms six out of the 11 target landscapes were present. The length of time in agri-environment schemes varied from 1 to 8 years.

The financial impact of agri-environment scheme participation pre-decoupling

The financial impact of applying management prescriptions on the three farms is set out in Table 18.2. This shows that all of the farms benefited financially under the management prescriptions chosen, with the exception of a small loss made under WES by Farm B. However the scale of the benefit differed considerably both farm-by-farm and prescription-by-prescription. Overall Farm C received the highest payment but also incurred the greatest income foregone. Farm A received the least payment but achieved the highest net financial benefit. Farm B achieved a similar net gain to Farm A but from more than double the level of payment.

The reasons for the differing results are explored in detail in Jones (2006). They are often very specific to the farm in question and once they are investigated it can be appreciated why circumstances do not necessarily replicate those envisaged in the notional budgets on which the payment levels were based. It demonstrates why it is unlikely that any nationally determined estimate of income foregone would necessarily be replicated at farm level. The reasons are not simply to do with lower than expected loss in revenue. It is also to do with cost structures and in particular the use of contractors and the influence this has on marginal cost reduction in response to changes in cropping. This is a theme developed more fully in Jones (2005). It must be borne in mind that agri-environment prescriptions are typically applied to field margins and relatively unproductive land. Therefore conditions will not necessarily be representative of the rest of the field let alone the rest of the farm or the locality. It is also likely that farmers will select options that involve relatively little cost or change in management i.e. the issue of 'additionality' developed by CRER and CJC (2002).

The evidence from the three case studies is that lower income foregone can (and often does) come from combinations of lower income loss, greater cost savings, lower additional costs and greater income under the new land use. The circumstances that can give rise to this are summarized in Table 18.3. This could form the basis for a typology of what to look for in calibrating payment levels to actual income foregone rather than basing it on notional forecasts. This more precise calibration would avoid additionality and allow better targeting and appropriate uptake.

Table 18.2: The financial impact of the application of agri-environment management prescriptions pre- and post-decoupling

Prescription	Payment	Pre-decoupling Income foregone	Pre-decoupling Net gain or loss	Post decoupling Income foregone	Post decoupling Net gain or loss	Difference in net gain/loss due to decoupling
Farm A						
CSS 6 & 2 m margins	£3,335	-£1,972	£1,363	-£1,065	£2,270	£907
WES 6 m buffer strip	£1,212	£0	£1,212	£0	£1,212	£0
CSS educational access	£700	£0	£700	£0	£700	£0
CSS arable reversion	£1,344	-£570	£774	£475	£1,819	£1,045
CSS extensification of grazed pasture	£558	-£192	£365	-£192	£365	£0
Total	**£7,148**	**-£2,734**	**£4,414**	**-£783**	**£6,366**	**£1,952**
Farm B						
CSS 6 m margins	£2,748	-£1,793	£955	-£637	£2,111	£1,156
CSS buffer strips	£70	£0	£70	£0	£70	£0
CSS overwintered stubbles	£2,045	£0	£2,045	£0	£2,045	£0
CSS arable reversion	£1,710	-£1,234	£476	-£75	£1,635	£1,159
WES scrub clearance & winter grazing	£8,620	-£8,773	-£153	-£8,773	-£153	£0
CSS hedgerow management	£0	£750	£750	£750	£750	£0
Total	**£15,192**	**-£11,050**	**£4,142**	**-£8,735**	**£6,457**	**£2,315**
Farm C						
CSS 6 m margins	£1,099	-£753	£346	-£303	£796	£450
CSS 2 m margins	£3,776	-£3,083	£693	-£1,020	£2,756	£2,062
CSS 6 m buffer strips	£726	-£558	£168	£1,070	£1,796	£1,628
CSS beetle banks	£168	-£133	£35	-£72	£96	£61
CSS stone curlew plots	£2,940	-£2,780	£160	-£1,469	£1,471	£1,311
CSS arable reversion	£5,375	-£5,390	-£14	-£4,082	£1,293	£1,308
CSS extensive grazing	£1,496	-£2,357	-£860	-£842	£654	£1,515
WES extensive grazing	£725	-£644	£81	-£230	£495	£415
Total	**£16,306**	**-£15,699**	**£607**	**-£6,949**	**£9,357**	**£8,750**

Table 18.3: A summary of factors that can have a positive or negative impact on income foregone subject to on farm conditions and the type of agri-environment prescription applied

Disadvantages of entering the scheme	Advantages of entering the scheme
Income loss	**Income gain**
High: High gross margin enterprises (good field conditions for crop growth and workability and high value rotations)	**High:** Supplementary payments; good gross margins on new land use; triggers extensification top up payment on beef subsidies; quota leasing
Low: Low gross margins (due to flooding, salt spray, steep slopes, vermin damage, soil and drainage conditions etc.); uncropped land.	**Low:** No income gain other than the basic prescription payment itself (often the case)
Extra costs	**Cost savings**
High: Interest on extra working capital (especially beef cattle); need for beef or sheep quota leasing; extra labour; extra machinery; extra contract charges; housing requirements	**High:** Contract operations (e.g. combine harvesting, hedging); casual labour; interest on working capital savings (especially beef cattle); renting additional land for beef extensification payment
Low: Land is used for grazing agisted stock or topping only	**Low:** Marginal costs of mechanisation only (e.g. fuel, machinery repairs); labour cost is fixed or notional

Cost recovery on capital works

Capital works vary throughout the life of agri-environment schemes. Normally there is a tendency for expenditure to be higher at the start of a scheme than at the end particularly where this ties into the application of a management prescription, such as the fencing of land entered for arable reversion. It was only practicable in the study to look at the percentage of cost recovered on capital works in the year in which they were incurred. These results are shown in Table 18.4. There was a strong association between the cost of capital works and the use of contractors. The farm that showed a net gain (Farm C) did so because the activity (hedge planting) was undertaken by full-time labour in the middle of winter when effectively the opportunity cost of their input was zero and this fixed cost was not increased as a result of doing the hedge planting. In general terms payment levels on capital works were insufficient to cover the cost of the work. There might be some monetary and non-monetary benefit to the enhancement of the environmental and landscape infrastructure of the farm but this was not included.

Table 18.4: Grant payments and actual costs of capital works under both CSS and WES schemes on case study farms

Farm	Grant	Actual costs	Net gain/loss	Grant as a % of cost
Farm A	£3,257	£8,409	**-£5,152**	38.7%
Farm B	£2,192	£2,333	**-£141**	93.9%
Farm C	£3,620	£1,828	**£1,792**	198.0%

The impact of decoupling *ceteris paribus*

The impact of decoupling on the income foregone calculations on a *ceteris paribus* basis is also shown in Table 18.2. All three farms show a greater financial advantage from participation in the agri-environment schemes post-decoupling. Indeed if payments seemed high against actual levels of income foregone before decoupling this was even more apparent afterwards. But the farms with already low income foregone had less scope to gain by this. As a result the order of net financial benefit is dramatically reversed from the pre-decoupling position with Farm C now benefiting the most and Farm A the least. The changes vary according to the prescription as well as the on-farm circumstances. The detail of these changes is explored fully in Jones (2006). However in general terms the higher the subsidy intensity of the farm system pre-decoupling the more subsidy removal impacts the budget. Farm C had a farm system that was a major recipient of Beef Suckler Cow Premium, Herbage Seed Aid and Arable Area Payments. Farm A had a smaller stake in the previous subsidy system with its emphasis on unsubsidized arable crops. Farm C had sheep as well as cattle and as breeding sheep were less well supported than breeding beef cattle under the old support regime this affected the scope for improvement. The differences in subsidy concentration pre-decoupling linked to the differences in farming system are the main reason for the reversing of the order of net benefit. But another factor was that Farms A and B both benefited from activities which lay outside the remit of agricultural subsidies (educational access and hedge cutting). Naturally enough these were unaffected by subsidy changes.

It is worth noting that prior to decoupling hedge cutting was the only activity to show a positive income foregone, i.e. there was a financial benefit to applying the prescription even without the payment. This came about because it was cheaper to cut hedges less frequently. But after decoupling other management prescriptions show themselves capable of showing a positive income foregone, i.e. the activity made a loss so stopping it increased rather than decreased income. These examples were arable reversion on Farm A and 6 m buffer strips on Farm B. This shows that some systems heavily reliant on subsidy for profitability are not going to represent a reasonable starting point for analysing change post-decoupling.

Conclusions and implications

The case study assessment offers some very clear conclusions and some important implications for the use of income foregone based agri-environment payments as an instrument of policy. While the evidence is based on only three farms it is clear from this evidence that even within broadly similar farm type categories substantial farm level variation in costs and income foregone exists. Furthermore it proves that this variation in the cost-benefit from agri-environment scheme participation can be observed for the same prescription, under the same agri-environment scheme, at the same rate of payment but on different farms.

In general there was a net financial benefit on all three farms (based on the budgeting basis chosen). Participation in agri-environment schemes thus increased farm income despite the fact that the payment rates were set with little or no allowance for any profit or 'incentive' element. This might lead one to conclude that the budgets used to justify the payment rates overstated the costs and income foregone. However this does not necessarily follow. The budgeted figures have to be based on 'normal conditions'. In practice the financial losses from implementing an agri-environment prescription can vary substantially around this norm (as shown by the results from the case study farms). Farmers might be expected, on economic grounds alone, only to enter schemes when they show a worthwhile profit. They have a degree of choice over which agri-environment prescriptions to join. As a result there is bound to be a bias towards situations where the costs and income foregone of participation are particularly low in survey work. It would obviously be necessary to survey farms that are not in agri-environment schemes to prove this point conclusively. But it seems logical and appears to be born out by the fieldwork.

It is possible to identify the factors (in Table 18.3) which create situations under which farmers might gain or lose by agri-environment scheme participation (thereby affecting their economic sustainability as well as the ecological sustainability of the wildlife on their farms). The results show that farms run by contractors generally stand to gain more by way of cost reductions as a result of a cutback in production under agri-environment prescriptions, than farms run by farming families without outside help. This has significant potential implications for the social sustainability of farming communities. However it is difficult to see how factors that have more to do with business cost structures than physical factors (such as field size or the productive capacity of the land) can be used to alter payment levels to suit the circumstances. There are difficult issues tied up in this. As policy moves its emphasis from Pillar I to Pillar II of the CAP, the dilemma about potential conflict between different sustainability criteria under Pillar II support measures will become more pertinent.

Taking account of the decoupling of direct payments from production had a dramatic impact on the extent to which the case study farms might be expected to benefit from agri-environment participation. The greatest beneficiaries were those farms that had received the largest amounts of subsidy under the old support regimes. As a result although all three farms showed an increase in financial benefit the order was reversed as to which benefited the most. Clearly payment rates may be expected to change. But even so this has important implications for farms that might not have wanted to enter agri-environment schemes in the past because of the loss of production linked subsidies, but might now be encouraged to do so. It must be remembered however that there are some prescriptions that are totally unaffected by the subsidy shift.

The case study results illustrate how vital it is to look in detail at farm level impact to understand the process of why farmers enter agri-environment schemes and why they choose the options that they do within them. This has become even more important in the context of a more substantial shift of funds from Pillar I to Pillar II and the introduction of open access non-competitive agri-environment schemes such as Entry Level Stewardship (ELS). Under ELS farmers have a free choice of which prescriptions to opt for and as long as this meets the points target they cannot be refused entry. In these circumstances a better understanding of farm level cost-benefits is even more vital to appreciating the role that agri-environment will play within sustainable systems of farm management in the future.

Acknowledgements

The author is very grateful to the farmers of the three case study farms and the RSPB who funded the original survey work. Thanks are also due to Paul Silcock (now of Cumulus) who was the project team leader for GFA/RACE in the work for the RSPB.

References

Council of the European Union (1999) Council Regulation (EC) No. 1257/1999 of 17th May 1999 on support for rural development from the EAGGF and amending and repealing certain regulations. *Official Journal of the European Union* L 160, 42, 26.6.1999, 80-101.

Council of the European Union (2002) Council Regulation (EC) No. 445/2002, of 26th February 2002 laying down detailed rules for the application of Council Regulation No. 1257/1999 on support for rural development from the EAGGF. *Official Journal of the European Union* L 74, 45, 15.3.2002, 1-34.

Council of the European Union (2005) Council Regulation (EC) No. 1698/2005 of 20th September 2005 on support for rural development by the EAFRD. *Official Journal of the European Union* L 277, 48, 21.10.2005, 1-40.

CRER and CJC (2002) *Economic Evaluation of Agri-Environment Schemes*. Centre for Rural Economics Research, University of Cambridge and CJC Consulting, UK (commissioned as an economic report for Defra).

Defra (2003) *Decoupling Direct Payments in the UK: Some Key Research Findings*. Defra Economic Reports, Defra, London, UK. [Accessed Oct 2006] Available from http://statistics.defra.gov.uk/esg/reports/decoupling/dcplgsummary.PDF

Defra (2004a) *England Rural Development Programme 2000-2006* as approved by Commission Decision C (2000) 3003, 11th October 2000. [Accessed Mar 2005] Available from http://www.defra.gov.uk/erdp/national/default.htm

Defra (2004b) *Partial Regulatory Impact Assessment of Options for the Implementation of Cross Compliance - Good Agricultural and Environmental Condition*. Product Code PB9499, Defra Publications, London, UK.

Frost, A., Stewart, S., Kerr, D., MacDonald, J. and D'Arcy, B. (2004) Agricultural environmental management; case studies from theory to practice. *Water Science and Technology* 49(3) 71-79.

Impey, L. (2002) Biodiversity can inflict a host of hidden costs. *Farmers' Weekly* 15th February 2002, 62.

Jones, J.V.H. (2004) Oral evidence, written evidence and supplementary memorandum to written evidence in House of Commons Environment Food and Rural Affairs Committee *Implementation of CAP Reform in the UK*. Seventh Report of Session 2003-04, HC 226-II, The Stationary Office Ltd, London, UK.

Jones, J.V.H. (2005) To produce or not to produce - is that the question? *Journal of Farm Management* 12(5) 235-249.

Jones, J.V.H. (2006) A case study based assessment of the financial impact of agri-environment scheme participation; pre- and post-decoupling. *Journal of Farm Management* 12(8) 477-498.

Morris, J., Mills, J. and Crawford, I.M. (2000) Promoting farmer up-take of agri-environment schemes: the Countryside Stewardship Arable Options Scheme. *Land Use Policy* 17, 241-254.

Morris, C. and Potter, C. (1995) Recruiting the new conservationists: farmers' adoption of agri-environmental schemes in the UK. *Journal of Rural Studies* 11(1) 51-63.

Produce Studies (1994) *The CSS: A Review of Incentive Payments*. Unpublished report to the Countryside Commission, UK.

Savills (1994) *CSS: Review of Arable Reversion Incentive Rates*. Unpublished report to the Countryside Commission, UK.

Short, C., Winter, M., Crabb, J., Temple, M., Dauven, A. and Augustin, B. (2000) *Economic Evaluation of the Countryside Stewardship Scheme*, Report by CCRU and ADAS for MAFF. CCRU, University of Gloucestershire, Cheltenham, UK.

Tate, G.J. (2001) *An Evaluation of the Shropshire Hills ESA from the Farm Business Perspective,* Unpublished PhD Thesis, University of Reading, Reading, UK.

Thomson, V. and Slee, W. (1998) The financial impact of ESAs on LFA farmers - a case study of farmers in the Cairngorm Straths ESA. *Farm Management* 9(12) 618-633.

Turner, J.T. and Taylor, M. (1998) *Applied Farm Management*. Second edition, Blackwell, Oxford, UK.

Wilson, G.A. and Hart, K. (2001) Farmer participation in agri-environmental schemes: towards conservation-oriented thinking? *Sociologia Ruralis* 41(2) 254-274.

Part V

Political and Policy Frameworks

Chapter 19

International perspectives on sustainable farmland management

C. Potter

Introduction

The international policy context is coming to be seen as increasingly decisive in framing the issue of sustainable farmland management in the UK. This is true in both a technical and a broader philosophical sense. Technically speaking, changes to the structure of domestic agricultural support have already taken place in order to ensure it is more compatible with World Trade Organization (WTO) rules and an emerging international model of agri-environmental policy design. Despite the suspension of the Doha trade round in July 2006, countries around the world have embarked on a process of 'decoupling' the support farmers receive from their production decisions in order to lessen their trade distorting impact. This is reshaping the policy context in which farmers take decisions about farmland management and promises to bring about changes both to the pattern of land occupancy and the structure of land use. The vexed international debate surrounding agricultural subsidies has focused attention, amongst other things, on the effectiveness of agri-environmental measures and questions are being asked about how best to design schemes in order to ensure they are 'fit for purpose' in a more internationalized and fiscally constrained policy context. Meanwhile, a fundamental and politically rather fierce debate is underway about the environmental and social desirability of agricultural liberalization as an encompassing policy project. In agricultural terms, this focuses on the likely land use implications of the increased market access, reduced levels of state support and expansion in world trade that would be the logical consequences of any eventual WTO trade agreement. Rather than the sustainability of farmland management, it is the multifunctionality of agriculture that is seen to be at stake here, different positions being taken internationally and within the European Union on the extent to which this can and should be safeguarded under a more open international trading regime. The current chapter offers a brief overview of these inter-related features of the international agricultural policy debate. In discussing recent contributions the chapter looks first at some of the short term implications of changes to the structure of European farm support that have been put through during the course of the Doha round. It goes on to discuss the larger issues that are polarizing international debate and which remain unresolved in the wake of the recent breakdown of negotiations.

Farmland sustainability under a decoupled CAP

Most commentators now agree that the Agreement on Agriculture (AoA) which concluded the Uruguay trade talks in 1996 moved agricultural policy governance a step closer to the neoliberal model. As Peine and McMichael (2005) point out, the AoA requires adherence to liberalization as an informing policy principle. As a set of legally binding disciplines, the AoA subjects the agricultural policies of members to a set of rules and procedures designed in the long run to decouple domestic support, increase market access and reduce the use of export subsidies (the 'three pillars' of subsequent WTO agriculture negotiations). So far as domestic support is concerned, it defines a zone of compatibility of domestic subsidies with trade objectives known as the 'Green Box' and another transitional zone of lesser compatibility know as the 'Blue Box'. Throughout the Uruguay round and in the period following the AoA, many industrialized countries began to re-examine the structure of their domestic support in order to respect these newly enacted WTO disciplines. In the European Union, constraints on the use of export subsidies written into the AoA, together with an awareness of the need to adjust the Common Agricultural Policy (CAP) to prepare for eastward enlargement, led to the Agenda 2000 reforms of March 1999. These began to decouple farm policy by substituting arable areas and headage payments for (highly trade distorting) price support but as Blue Box payments they fell short of the conditions required for a full decoupling of payments and were soon superseded by the more radical Mid-Term Review (MTR) reforms of June 2003. Put in place as an earnest of the EU's commitment to the new negotiating mandate of the by now fully launched Doha trade round, these replaced direct payments with a (supposedly Green Box compatible) Single Payment (SP), available to farmers on an area basis and, notionally at least, decoupled from their decisions about how much to plant and how heavily to stock.

Importantly, so far as agricultural sustainability is concerned, the implementing regulations for the SP require all member states to impose environmental conditions on the receipt of support payments (known by its US nomenclature of 'cross compliance'). This is seen by supporters as a favourable change to the architecture of farm support because the majority of farmers now in receipt of these payments become subject to a more broad based form of environmental conditionality than has been possible through agri-environmental schemes alone. The EU signalled at this point its commitment to a combined strategy towards the promotion of agricultural sustainability in a post-Doha world, using both cross compliance and the deployment of more targeted incentives under agri-environmental programmes to encourage more sustainable land management on farms (CEC, 2003). But growing scrutiny from international trading partners within the context of the Doha negotiations meant that the case for environmental compensation for environmental management had to be more clearly specified than it had been in the past. Under the EU's Rural Development Regulation, therefore, it is laid down that agri-environmental measures must adhere to the requirement that farmers be required to meet conditions going beyond 'good agricultural practice'. As Rodgers (2004) observes, the result is that the law on the CAP now establishes good agricultural practice as a normative standard for the management of farmland across the EU.

It is still difficult to assess the implications of these linked policy reforms, many of them at this stage still essentially legalistic in nature, or the more substantive impact of the decoupling process, for farmland sustainability in the UK and across the EU. In technical policy design terms, the establishment of a baseline standard allowing policymakers to differentiate between actions that should be regulated without compensation and those that will attract subsidy is a welcome development. It clears the way for a more rigorous implementation of the polluter pays principle in agriculture than has been possible before and should make agri-environmental payments more transparent, and thus politically defensible. Much though depends on the willingness of agriculture departments to restrict payments to cases where an additional benefit over and above good agricultural practice can be proved. The current preference in many member states is for payments that merge environmental objectives with income support objectives (see further discussion of this stance on multifunctionality below), leading to situations where it is often difficult to prove empirically that any 'additionality' has been achieved (Hanley *et al.*, 1999; Kleijn *et al.*, 2006). A large component of an agri-environmental payment may therefore be for the maintenance of current (traditional) practices, making it difficult to establish the additional benefit that is being purchased through a scheme. Taken together with the uneven quality of agri-environmental monitoring and assessment procedures, and the elastic nature of the concept of good agricultural practice itself, the reality is somewhat different to that projected by European negotiators into WTO debates. Meanwhile, the relationship between cross compliance and agri-environmental payments remains complex. Dobbs and Pretty (2005) remark that so long as CAP payments remain high, cross compliance measures can effectively operate as baseline regulations for most farms (enforcing a concept of good agricultural condition on farmland). With reduced levels of expenditure on the SP, cross compliance loses its leverage, however, and the onus shifts back to some combination of regulation and agri-environmental support. The implication is that policymakers cannot for much longer avoid a debate about where precisely to draw the line between regulation and subsidization.

We know that any agricultural restructuring brought about by the decoupling process will have implications for sustainability given the close relationship between much of the natural environment and the activity of farming. Agricultural restructuring can be a powerful driver of change in the wider rural environment. One of the key sources of uncertainty in the short- to medium-term is how the SP itself will be seen by farmers and land managers. If it used to cross-subsidize other enterprises, we might expect a smaller movement in favour of extensification, de-stocking and withdrawal than many modelling exercises presently predict (see, for instance, GFA-RACE/IEEP, 2003). While there is expected to be some environmental gain due to an extensification of production brought about by decoupling, particularly of livestock subsidies, this needs to be set against intensification elsewhere due to the increased opportunities for agricultural exports of a more open market. For the moment, the evidence for the UK suggests that farmers are delaying change until they can assess what the new system of payments will mean for household income and cash flow (Lobley and Potter, 2004). The structural consequences of the new policy mix are hard to judge at this point, though they are unlikely to be neutral given an impact on land and asset values, and thus on long range expectations and entry and exit behaviour. It seems reasonable to expect the SP to act as a safety net for many farmers, further encouraging a delay in retirement in the case of elderly farmers, for instance, and thus increasing the number of retirement holdings in existence. In the dairy sector, on the other hand, the SP may well speed up a restructuring of agricultural enterprises to the extent that that remaining producers with smaller herds will be encouraged finally to leave dairying, switching to beef or just opting for the minimal management necessary to meet cross compliance requirements. In these situations there could be a significant rebalancing of enterprises, albeit within a reasonably stable set of farm structures. At the same time, the promised boost in funding for 'second pillar' rural development measures under the MTR and a stronger commitment to rural development generally will change the local economic environment in which farmers operate, facilitating a further diversification of the income base of many households, and thus enabling more of them to continue in their occupation of holdings, even if this requires more complicated land holding arrangements to do so.

Competing policy models within the WTO

The WTO process, then, is already reshaping the way farmland is managed, even without the further, more radical liberalization of agricultural markets and policies that many expected would conclude the latest round of negotiations. Despite the suspension of negotiations in July 2006, this larger project remains powerfully in play, supported by the corporate interests that increasingly comprise the global agri-food industry. The possibility of a much more extensive opening up of European agriculture to world market forces must therefore continue to be taken seriously by those interested in sustainable farmland management. Here we move beyond the merely technical policy design issue of how best to ensure stewardship measures remain within the legal terms of WTO trade rules, to consider the much larger question of the scale and pattern of farming activity that would be possible in a neoliberal policy setting. A central concern of debates which have taken place throughout the Doha round and through the work of organizations like the OECD (see, for instance, OECD, 2001) has been how best to ensure the continued provision of public environmental goods in a much more neoliberal context. There appear to be two competing policy models on offer here.

The first, aligned with the concept of multifunctionality, is generally pessimistic that sufficient farmers or levels of farming activity would be possible at world market prices to sustain the sort of managed countryside that European citizens appear to value and wish to see continue. A key assumption here is that nature conservation goals can best be achieved as a by-product of agricultural production. Hodge (2003) comments that one of the most persuasive arguments of 'multifunctionalists' is the degree to which human and natural systems have co-evolved in European (and other) settings, the maintenance of much of our semi-natural habitats and managed landscapes requiring the continuation of farming practices like grazing, mowing and burning in order to retain their biological and aesthetic value. This contrasts with many 'New World' systems, which are about sustaining 'pristine' environments apart from farming and where the emphasis on preventing soil erosion and off-site damage means that agricultural production and environmental protection are often regarded as inherently conflicting policy goals. The logical policy consequence of the multifunctionalist analysis is that large numbers of farmers need to be retained on the land in order to continue producing a managed countryside. This is an analysis that finds little favour with many of the EU's trading partners. Critics such as the Cairns Group and, to a more nuanced degree, the US, see multifunctionality as an excuse for continued high (and trade distorting) levels of agricultural support and have largely dismissed the concept as a guiding principle for future agricultural support (Cardwell *et al.*, 2003). The difficulty for advocates of a multifunctional model is being able to separate their interests in public goods from the rent-seeking motivations of a farm lobby anxious to defend their traditional policy entitlements. Multifunctionality arguments have been strongly seized on in the UK by the National Farmers' Union and the Country Landowners' Association, for instance, to justify current income support mechanisms. These organizations employ arguments broadly in line with those so influential in France, which say that only by protecting domestic farmers and agricultural businesses can we protect our rural environment. The status and defensibility of the SP is emerging as an important first test of these arguments. Is it an income support designed to keep the countryside occupied and managed by large numbers of family farmers or a simply another policy entitlement for farmers with no clear environmental rationale and regressive distributional consequences? Significantly, perhaps, the EU has chosen to emphasize its status as a decoupled income support rather than a payment that aims to procure a range of environmental outcomes via the conditions attached to it (CEC, 2003).

These definitional issues, and the political difficulties to which they give rise, have led many to explore an alternative public goods model to sustainable farmland management. This shifts the focus away from farmers and their occupancy of land towards the environmental outputs that need to be achieved. As advocated in the UK (HM Treasury and Defra, 2005), this takes a more sanguine view of farmers' ability to adapt to changing market conditions by pointing to the evidence of a more diversified farm household income base and the trend towards pluriactivity and off-farm employment. Moreover, under this approach it is not deemed to be the responsibility of the state to fashion agricultural land use more generally. As the Treasury paper puts it:

> it will not be an objective of the new CAP to maintain existing or specific levels or patterns of production, whether within individual Member States or across the EU as a whole. Rather, production should be allowed to find a more sustainable level, reflecting natural advantages (in terms of climate and terrain), competitive advantages (in terms of food quality and safety) and rational trading relationships in a more open market (HM Treasury and Defra, 2005: 15).

Environmental outputs, meanwhile, are viewed as entirely separable from the activity of farming and thus capable of being paid for and secured under a (presumably extensive) system of agri-environmental contracts and private agreements which pay by results and outcomes. Largely in line with the decoupled model favoured by the WTO, such arguments are much more to the liking of trading partners. However, questions remain about the delivery mechanisms available and the degree to which these can be made fit for purpose. Administrative costs tend to be high and there are difficulties in targeting and delivering geographically differentiated outcomes. At the same time there is concern about the sustainability of a set of landscapes that would increasingly depend on government schemes for their existence and construction. Far better, and more authentic, say the multifunctionalists, to retain the idea of farmed landscapes as working environments, underpinned and sustained by the activities of the individuals and families seeking to gain a livelihood there.

Conclusions

The international context is becoming increasingly important in any discussion of farmland sustainability. An emerging infrastructure of trade rules, subsidy codes and dispute procedures means that the governance of farming activities through public policy is more and more being decided in forums like the WTO. The AoA already sets boundaries on what national governments can do in terms of offering subsidies to farmers and we have seen changes to the pattern of farm support in the EU that have been a direct response to the politics of the Doha negotiations. It seems unlikely that the trend towards decoupling subsidies from production and 'recoupling' them to environmental and rural development goals will not continue. There remain fundamental differences of opinion regarding how far this process should be taken, however, internationally and within the EU. Rather than being a debate about farmland sustainability as such, the focus of international discussion in Geneva, Brussels and Washington has been how far the multifunctionality of European agriculture could be put at risk by a long term restructuring of farming due to increased world market exposure. For many, this debate has a causal emphasis which the often rather vaguely prescribed notion of sustainability lacks. It draws attention to the difficulty of reconciling globalizing tendencies with local attributes and processes and has offered something of a counter to the dominant narrative of market rule as a governing discourse. The question of how far governments have a responsibility to sustain the countryside by protecting the incomes of farmers is still one that remains unresolved after over a decade of contestation and debate.

References

Cardwell, M., Grossman, M. and Rodgers, C. (eds) (2003) *Agriculture and International Trade: Law, Policy and the WTO*. CAB International, Wallingford, UK.

CEC (2003) *Long Term Policy Perspective for Sustainable Agriculture*. Commission of the European Communities, Brussels, Belgium.

Dobbs, T. and Pretty, J. (2005) Agri-environmental stewardship schemes and multifunctionality. *Review of Agricultural Economics* 26(2) 220-237.

GFA-RACE/IEEP (2003) *A Long-Term Policy Perspective for Sustainable Agriculture: The Potential Environmental Impacts of CAP Mid Term Review Proposals*. Research Report for Defra. Defra, London, UK.

Hanley, N., Whitby, M. and Simpson, I. (1999) Assessing the success of agri-environmental policy in the UK. *Land Use Policy* 16, 67-80.

HM Treasury (HMT) and Defra (2005) *A Vision for the Common Agricultural Policy*. HMT and Defra, London, UK.

Hodge, I. (2003) Mainstreaming rural development policy under the CAP: an English perspective. *Journal of Environmental Policy and Planning* 5(4) 361-378.

Kleijn, D., Baquero, R.A., Clough, Y., Díaz, M., De Esteban, J., Fernández, F., Gabriel, D., Herzog, F., Holzschuh, A., Jöhl, R., Knop, E., Kruess, A., Marshall, E.J.P., Steffan-Dewenter, I., Tscharntke, T., Verhulst, J., West, T.M. and Yela, J.L. (2006) Mixed biodiversity benefits of agri-environmental schemes in five European countries, *Ecology Letters* 9(3) 243-254.

Lobley, M. and Potter, C. (2004) Agricultural change and restructuring: recent evidence from a survey of agricultural households in England. *Journal of Rural Studies* 20, 499-510.

OECD (2001) *Multifunctionality: Towards an Analytical Framework*. OECD, Paris, France.

Peine, E. and McMichael, P. (2005) Globalization and global governance. In: Higgins, V. and Lawrence, G. (eds) *Agricultural Governance: Globalization and the New Politics of Regulation*. Routledge, London, UK, 19-34.

Rodgers, C. (2004) Environmental policy and the reform of European agricultural law. In: Cardwell, M., Grossman, M. and Rodgers, C. (eds) *Agriculture and International Trade*. CAB International, Wallingford, UK, 277-300.

Chapter 20

The national policy dimension for environmentally sustainable agriculture: a UK perspective

G. Morgan and C. Reid

Introduction

Since joining the European Common Market in 1973, and thus becoming subject to the Common Agricultural Policy (CAP), the overall policy framework for agriculture in the UK has been overwhelmingly established at the European level. UK governments have subsequently sought to influence the CAP to advance both national objectives for agriculture and also their wider policy agenda. Within the UK, the implementation of agricultural policy and the use of agricultural policy tools have become increasingly a devolved matter. Policy objectives in England have been set out in the *Strategy for Sustainable Farming and Food* (SSFF) (Defra, 2002), which was stimulated by the challenge facing the industry following the major outbreak of foot and mouth disease in 2001 and the identification of Bovine Spongiform Encephalopathy (BSE) in humans in 1996. The SSFF identified a need to change the nature of agricultural subsidies and highlighted the degree of trade distortion they caused, and also discussed the consequential impact on Less Economically Developed Countries (LEDCs), domestic prices and the environment. This suggests that domestic policies for sustainable agriculture are nested within, and potentially compete with, other important policy domains: not only competitiveness and trade liberalization, but also 'better regulation' (Defra, 2005a), sustainable development (HMG, 2005), climate change strategy and EU environmental policy (e.g. the Habitats Directive and the Water Framework Directive). Furthermore, agricultural support increasingly needs to compete for resources with other more eye-catching priorities for public expenditure.

The most recent UK policy position on public support for agriculture is found in the long-term *Vision for the CAP* produced jointly by Defra and the Treasury in 2005. This document presents a trenchant critique of the impact of current support policies and proposes that 'a sustainable CAP would comprise [*inter alia*]:

- a clear framework, set at EU level, to define the goals of EU agricultural policy, focussing in particular on maintaining the

environment and promoting sustainable rural development, particularly in the more environmentally sensitive areas of the EU;
- EU spending on agriculture ... based on the current Pillar II ... allowing a considerable reduction in spending by the EU on agriculture (HMT and Defra, 2005: 4).

Although it thus appears that improving environmental quality is a major driver of agriculture policy, it is open to question how far such rhetoric on environmentally sustainable agriculture is underpinned through national policy implementation and in EU negotiating positions. To illuminate this issue, we examine the extent to which environmental priorities are reflected in the UK approach to a series of agriculture policy instruments and objectives, and conclude by suggesting that further progress in this direction is by no means assured.

Agricultural liberalization and the decoupling of farm subsidies

The UK Government wishes to secure an agricultural industry that is 'internationally competitive without reliance on subsidy or protection' (HM Treasury and Defra, 2005: 9). This would comprise 'a free, fair and level playing field' for farmers, with EU agriculture spending based on the current Pillar II (HM Treasury and Defra, 2005: 16). The decoupling of farm payments from production heralded in the 2003 Mid-Term Review of the CAP (Council of the EU, 2003) represented a major step in the direction of reform favoured by the UK, continuing a process that started with the onset of area payments in 1992 and continued in the Agenda 2000 reforms. Unsurprisingly, the UK opted for complete decoupling from 2005, whilst a number of other Member States exercised the option of retaining a degree of coupling. However, the constituent UK countries adopted different ways of paying the new Single Payment (SP) that replaced the former support regimes, with Scotland and Wales using the default historic basis, England phasing in an area payment over time and Northern Ireland using a hybrid approach.

The 2003 reforms were welcomed by some organizations - for example the Royal Society for the Protection of Birds (RSPB) described the deal as a 'substantial breakthrough' for the environment (RSPB, 2005a: 8) - because decoupling was expected to help address some of the problems associated with the intensification encouraged by production subsidies such as over-grazing of heather moorland by sheep (JNCC, 2002; GFA-RACE/IEEP, 2003). However, it cannot be assumed that for the UK government the environment was an overwhelming imperative for agreeing to the political deal at the heart of the reforms. The *Vision for the CAP* (HM Treasury and Defra, 2005) suggests that other drivers, including trade liberalization and the desire for a more market-based industry, continue to be a significant influence; and the subsequent EU budget deal for the period 2007-2012, brokered by the UK acting as EU President, disappointed many by sidelining the original Commission proposal for a substantial increase in EU resources for environmental agriculture. Indeed, in response to this action the RSPB put forward the view that 'Europe's wildlife has been made a scapegoat, sacrificed so that the UK can push through its controversial budget proposals' (RSPB, 2005b).

The 2003 deal also failed to address the issue of cutting the aggregate level of support for farming, leaving the UK with a quandary over the operation of the new SP scheme. The purpose of the scheme was left unclear in the 2003 deal, although in the new Rural

Development Regulation for 2007-2013 (Council of the EU, 2005) the European Commission referred to it as a basic income support for farmers. However, across the EU as a whole the payment largely reflects historic receipts; and since in 2000, 1.26% of farmers received 25.4% of CAP direct payments with 5.3% of farmers receiving 50.2% of direct payments (Eurostat, 2002), the new SP scheme will be heavily skewed towards larger farms. This issue was mirrored in the UK where in 2000 81% of Pillar 1 payments went to 26% of farmers (Eurostat, 2002), leading the UK to reject explicitly the logic of such an income payment in its *Vision for the CAP* (HM Treasury and Defra, 2005: 29).

Indeed, the lack of any clear rationale for the SP scheme as a policy instrument in promoting sustainable farming appears to have caused confusion in the UK. The redistribution and administrative complexity entailed in establishing an area-based Pillar I payment system in England was considerable, with the National Farmers Union (NFU) proclaiming itself 'dismayed' with the situation in 2006 (NFU, 2006). Furthermore, this was all undertaken against the backdrop of government opposition to the basic philosophy of Pillar I payments. Thus the Secretary of State for Environment, Food and Rural Affairs, Margaret Beckett, announced early on that, due to 'accompanying complexity and loss of transparency', she did 'not propose to take advantage of the option for so called "National Envelope" measures in England' (Defra, 2004a: para. 19), which allow Member States to deduct a maximum of 10% of sectoral budgets to fund farming practices that lead to environmental protection or enhancement or the improvement of product marketing or quality (Council of the EU, 2003: 24). The situation also threatened to add to the political difficulty of further national modulation of Pillar I payments to fund agri-environment schemes under the CAP second pillar.

Agri-environment schemes

At the domestic level, although with significant variation across the UK, government has made considerable progress with the development of an alternative model for agricultural support. Agri-environment schemes, which provide payments to land managers prepared voluntarily to undertake environmental measures on their land, have been a cornerstone of UK environmental policy since 1986 (Dobbs and Pretty, 2005). They are founded on the principle of paying public money (through a calculation of income forgone plus costs incurred) for environmental public goods that would otherwise be under-provided.

The Environmentally Sensitive Areas (ESA) scheme which was piloted and run by the Ministry of Agriculture Fisheries and Food (MAFF), and the Countryside Stewardship scheme initiated by the Countryside Commission, became the forerunners of European-wide agri-environment policy which became embedded in the McSharry reforms of the CAP in 1992 (Lowe *et al.*, 2002). The early ESA schemes were primarily designed to protect environmentally sensitive areas at risk from agricultural intensification, with later schemes such as the Habitats scheme and the Countryside Stewardship scheme extended to include restoration and re-creation of habitats and landscape features. These had often been damaged in the past by agricultural intensification, ironically often encouraged by the implementation of government policies directed towards food security, such as grants for land drainage and hedgerow removal. However, agri-environment scheme funding was dwarfed by the environmental challenge of agricultural intensification impelled by CAP production support.

The UK broadly supported the Agenda 2000 CAP reforms which formalized the creation of a 'second pillar' within the CAP, directed towards rural development and agri-environment payments. A major review of agri-environment schemes took place in England from 2001-2004, prompted in part by the recommendations of the Policy Commission under Sir Donald Curry, set up in the wake of the foot and mouth epidemic (Cabinet Office, 2002), and also by the possibility of switching EU agricultural payments into agri-environment schemes. Environmental Stewardship (ES), the new English scheme which emerged to replace existing schemes, was designed to be more tightly focused on achieving the government's environmental targets, many of which were by now more clearly articulated (e.g. through the UK Biodiversity Action Plans and in governmental Public Service Agreement targets for farmland birds and Sites of Special Scientific Interest - SSSIs). Environmental Stewardship included an 'Entry Level' (ELS) for widespread, but relatively small-scale, improvements across the whole farmed landscape; and a 'Higher Level' (HLS) for more demanding management of particular habitats and features. The *Strategy for Sustainable Farming and Food* committed significant levels of scheme funding through switching Pillar I payments into the England Rural Development Programme and match-funding these with Treasury resources (Defra, 2002: 27). The new agri-environment schemes generated considerable optimism about the prospects for more environmentally sustainable agriculture, particularly when viewed alongside improvements in flanking policies driven from the EU level, such as the introduction of Environmental Impact Assessment for uncultivated land and the decoupling of Pillar I payments from production. However, the environmental success of the new agri-environment schemes remained dependent on identifying appropriate objectives and options, and ensuring sufficiently widespread uptake; and also upon sustaining the practices the schemes promote.

Environmental Stewardship contained options addressing a wide range of environmental objectives - including water and soil quality, biodiversity maintenance and enhancement, conserving landscapes and the historic environment, and countryside access. In addition, the EU strategic guidelines for rural development (Council of the EU, 2006) contained encouragement for future schemes to address climate change mitigation and adaptation. It has been demonstrated that agri-environment schemes can restore species populations, recover damaged habitats and improve ecological water quality, thus making land management more sustainable (Pywell *et al.,* 2006). However, schemes generally do not address environmental issues at the farming systems level (other than those offering support for organic farming practices), so that environmentally harmful practices can still be taking place on a holding even though some of the land area is subject to an agri-environment scheme.

The Entry Level of Environmental Stewardship allowed for widespread uptake of scheme measures at a low level on each holding - the government estimated up to 60% of holdings would be in the scheme by the end of 2007 (Commons Hansard, 2006: col. 243W). The Higher Level aimed to be more targeted towards areas of greatest environmental need, and to move away from the rather scattergun approach of the earlier Countryside Stewardship scheme. However, although scheme funds have increased over the last 7 years, resources are still limited in relation to targets for habitat restoration and creation and improvements in catchment management. Furthermore resources cover an increasing range of scheme objectives, suggesting that uptake of higher level options across the countryside may be insufficient to achieve key environmental objectives over the next few years.

There is an assumption that land managers revert to previous - and sometimes environmentally damaging - practices in the absence of agri-environment payments, with schemes often paying for specific practices that are not necessarily well integrated with other aspects of the farm business. However, the new EU rural development policy encourages greater integration between the environmental, social and economic elements of sustainable rural development, encouraging land managers to capitalize on the economic benefits that a high quality environment can bring (e.g. to boost the tourism potential of their land or to market products on the basis of the high quality environment they are derived from), on the basis that the more integral environmental land management becomes to businesses and communities, the more likely it is to be sustained over the long term. The UK's *Vision for the CAP* makes clear that the government sees a long-term role for agri-environment schemes to deliver environmental public goods (HM Treasury and Defra, 2005: 18).

Future developments in agri-environment approaches in the UK might include tendering or bidding systems; community or group designed schemes (e.g. using the Leader approach under the European Agricultural Fund for Rural Development (EAFRD) Regulation - Council of the EU, 2005); and cooperative management of the agricultural environment. Such approaches could be used to help address emerging issues for sustainable agriculture, which need to be tackled at a scale wider than the individual holding - such as mitigation and habitat adaptation to climate change; making links between a healthy natural environment and local marketing initiatives; larger scale afforestation; conservation grazing schemes; and diffuse water pollution from agriculture. However, with many elements of scheme design and function closely controlled at EU level - such as the income-foregone method of calculating payments and limitations on those eligible for funding - there may be an increasing tension between national and European policy aspirations.

Additional Pillar I policy instruments

Although UK agriculture policy is most sharply illuminated by the issues of liberalization and support for provision of environmental public goods, the approach taken to the application of other CAP policy instruments throws further light on the UK approach to environmentally sustainable agriculture.

Cross compliance

Agenda 2000 introduced the possibility of EU Member States making certain farm payments subject to cross compliance, i.e. making farm payments conditional on specific (usually environmental) requirements or subject to compliance with other legislation (such as animal welfare regulations). In the UK these were translated into basic conditions attached to set-aside management and to agri-environment and Less Favoured Area (LFA) payments (Dwyer *et al.,* 2000: 25).

The CAP reforms agreed in 2003 hugely increased the scope and applicability of cross compliance, requiring Member States to make receipt of the new SP subject to compliance with Statutory Management Requirements derived from 18 EU Directives. An additional requirement was to ensure the maintenance of 'good agricultural and environmental condition' defined at national level within an EU framework; measures to achieve this condition are mostly concerned with basic soil management issues, along with a requirement to ensure 'a minimum level of maintenance' (Council of the EU, 2003: 58). It is believed that the proposal to amend the original title of 'good agricultural condition' to embrace the environment came from the UK.

The UK government describes cross compliance as one of those 'aspects (or by-products) of Pillar I of the CAP that are regarded as being environmentally positive. Whilst such benefits are clearly of value, they are not delivered in the most efficient and targeted way' (HM Treasury and Defra, 2005: 33). Cross compliance thus occupies a somewhat ambiguous position as a national policy instrument. Since the standards derived from Directives should already be enshrined in law, cross compliance makes no difference to those existing standards - although it does increase the risk and degree of sanction.[1] However, the requirement in Annex IV of the Regulation to make the SP subject to a minimum level of management is more problematic since it does not relate to existing requirements and might be seen as demanding positive environmental delivery. Thus some elements of the framework for cross compliance the UK agreed to in the 2003 reforms have created problems in implementation. Following the 2003 reforms, Defra, working closely with its environmental agencies, such as English Nature (EN) and the Environment Agency, adopted an approach to cross compliance which focused on internalizing environmental externalities, taking a much more ambitious approach to that taken after Agenda 2000. Thus although there is now a programme of measures under cross compliance to help address the issue of soil management, and Scotland has complied with the required introduction of a minimum level of maintenance for habitats (such as species-rich grasslands) (SEERAD, 2004), this is still awaited in England, being both technically challenging, and potentially sitting uncomfortably with the UK central government approach to CAP reform.

Cross compliance therefore creates tension between European and national policy approaches. Although in one sense it is a 'cheap' instrument to reinforce national standards, it threatens to antagonize farmers, imperilling payments regarded as 'entitlements'. Most of the reductions in farm payments resulting from infractions are returned to Europe, with a knock-on increase to the net national contribution to the EU. Furthermore the deregulation agenda resists the imposition of onerous standards fulfilling the requirement for minimum levels of maintenance. Most fundamentally, the UK position that Pillar I payments have no long-term legitimacy militates against securing long-term benefits through attaching environmental conditions to those payments. These issues were exemplified in England in the proposed requirement to create protective buffers for hedges and watercourses to fulfil the EU requirement to protect landscape features (Defra, 2006a). This was bitterly resisted by the industry (see, for example, Tenant Farmers Association (TFA), 2004) on the basis of imposing excessive administrative burden in relation to environmental benefit, and the resultant agreement (whereby hedges around smaller fields were exempt from the requirement) represented an uneasy compromise reflecting lack of clarity about the role of the policy instrument.

As Pillar I payments start to decline in real terms and the purchase of the SP on agricultural practice recedes, the acid test for English agri-environmental policy will be whether cross compliance standards are lost, or shifted into regulatory approaches; or alternatively whether they are used to justify the SP scheme as some kind of 'green' payment.

Set-aside

Compulsory set-aside was introduced in 1992 as part of the MacSharry reforms to help manage the impact on agricultural production of the shift from price support towards area payments. The requirement, attached to the former Arable Area Payment scheme and subsequently the SP scheme, made the payments subject to the removal from production of a variable proportion (up to 15%) of the arable area each year. The retention of set-aside as a market management tool in the decoupling reforms of 2003 seemed somewhat anomalous, suggesting that a significant reorientation to world markets was not in practice anticipated.

A greater degree of environmental conditionality for set-aside followed the Agenda 2000 reforms (e.g. limits on timing of herbicide applications), and there were further possibilities for undertaking beneficial environmental management (although without additional payment). Whilst there are undoubted biodiversity benefits of set-aside in the UK (for example in respect of breeding farmland birds and arable plants) partly as a result of the scale of the policy, the overall benefits are unclear (Firbank *et al.*, 2003). Some opportunities have been taken by Defra to improve the environmental performance of set-aside, such as introducing the possibility of narrower strips of permanent set-aside (from a minimum width of 10 m to 6 m) (Defra, 2005b), which could improve its use to reduce diffuse pollution impacts; but the basic controls over management of rotational set-aside remain largely unchanged with spraying of regenerated vegetation remaining the preferred agronomic management for the majority of such set-aside.

Although the UK government views set-aside as having 'potential benefits in respect of diffuse water pollution reduction and habitat creation', noting that these 'might otherwise have to be paid for through agri-environment schemes' (HM Treasury and Defra, 2005: 33), the overwhelming priority for the UK in the CAP 'Health Check' of 2008 is likely to be securing abolition of set-aside in pursuit of greater market liberalization. Thus, as with cross compliance, the ambiguous nature of set-aside creates a tension between the introduction of meaningful environmental standards and the long-term expectation that the instrument will disappear.

National envelopes

Agenda 2000 introduced the possibility of 'enveloping' a proportion of certain direct aids and using the resources generated to incentivize particular production systems in each sector. This was a potentially attractive instrument for the UK in that it allowed for a parallel approach to that of securing environmental benefits through national modulation, whilst not requiring national match-funding and potentially entailing less resistance from the industry since resources were only redistributed within a sector. In England, a 'sheep envelope' was created from sheep premium payments and administered by English Nature to secure reductions in grazing numbers on overgrazed uplands and the reintroduction of grazing on undergrazed (mostly lowland) sites (EN, 2003). This resulted in some significant environmental gains. For example, through sheep envelope agreements costing £5.5m, around 41,000 ha of nationally important wildlife site land (SSSI) was brought into sustainable grazing regimes. The UK played an important role in securing a similar mechanism in the 2003 reform agreement, and envelopes were subsequently adopted in Scotland and Northern Ireland. However, the decision not to continue use of the envelope instrument (Defra, 2004a) reflected both the problem of enveloping resources within an area-based SP scheme and the distributional impacts of the establishment of a complex new payment system which militated against the possibility of any further redistribution of CAP subsidies.

Support for farming systems

In general, financial support for agriculture in the UK has been channelled either via commodity intervention (through the CAP Pillar I) or towards particular management practices (such as agri-environment under Pillar II). However, there are two contrasting examples where support has instead been offered for particular farming systems.

Organic farming[2]

Organic farming is a whole system approach, with a prescribed set of standards which are monitored by professional certification bodies. The system eliminates use of artificial fertilizers and pesticides and promotes high standards of animal welfare, limiting the use of artificial medicines and treatments. The area of land farmed organically in the UK increased steadily into the early 21st century, with organic and in-conversion land in the UK totalling 619,783 ha at January 2007 (Defra, 2007a) and the retail market for organic products in the UK worth an estimated £1.213 billion in 2004 (Soil Association, 2005). This represented an approximately tenfold increase on 1994 figures, with encouragement to convert to organic agriculture coming from government policies (such as organic conversion payments and the new Organic Entry Level scheme (2005), which provides ongoing support for organic farmers). However, the extent of organic land has declined slightly in recent years (Defra, 2007b) and it still constitutes only a small proportion of the total agricultural area of the UK (3.6% in January 2007 - Defra, 2007a). Thus questions remain about the role of organic farming within the move towards more environmentally sustainable agriculture.

It is evident from a range of research (Bengtsson *et al.*, 2005; Hole *et al.*, 2005) that biodiversity and other environmental benefits routinely accrue from farms switching from non-organic to organic production. Other systems like Integrated Farm Management (IFM) also have practices that are beneficial to the environment (e.g. over-winter stubbles, precision fertilizer applications), but, unlike organic, these other systems do not currently have a regular, reliable and consistent certification standard, meaning that although they promote certain beneficial practices they are not necessarily implemented consistently or across the whole farm.

Organic yields are, however, typically less than yields from conventional farming. This poses a dilemma for government: whether in a world of increasing demand for food, it is appropriate to extend coverage of a more environmentally productive, yet lower yielding, system which could lead to destruction of natural habitat to increase the farmed area and potentially food shortages across the globe. Thus government policies to substantially increase the coverage of organic systems (and the attendant *in situ* benefits they provide for wildlife, water and soil quality, carbon and climate change) would have to be accompanied by wider policy changes of varying political palatability, such as abolishing set-aside, reducing global meat consumption and introducing plant breeding programmes focused on increasing yields within organic systems.

The trend in the UK of offering greatest encouragement to farmers to convert agriculturally productive land to organic, because this results in the most marked extensification and benefit to wildlife (Bengtsson *et al.,* 2005), contrasts with the experience in some other EU countries, which faced with the prospect of land abandonment, favour organic as a system of agriculture most likely to retain the socio-economic benefits of agricultural production in marginal areas (CEC, 2004). Thus the wider policy picture in the EU is likely to exert some influence on the future development of this sector.

Less Favoured Areas and the Hill Farm Allowance[3]

European support for LFAs has been in place since 1975 and was designed to support farming operations in mountain and hill areas negatively affected by climate, altitude, topography and remoteness. Between 1976 and 1999 the LFA regulation was implemented in England under the Hill Livestock and Compensatory Allowance, the objective of which was to support the continuation of extensive livestock farming in the hills and uplands, thereby helping to maintain a viable population and conserve the countryside by providing headage based payments to upland livestock farmers. The 1999 EU Rural Development Regulation revised the objectives for payments to farmers in LFAs identifying three key priorities:

- to ensure continued agricultural land use and thereby contribute to the maintenance of a viable rural community,
- to maintain countryside,
- to maintain and promote sustainable farming systems which in particular take account of environmental protection requirements (Council of the EU, 1999: 88).

Significantly the new regulation also required that compensatory allowances should be paid on an area basis, thus partially decoupling support. This was implemented in England as the Hill Farm Allowance (HFA) scheme in 2001. The new scheme also included additional 'environmental enhancements' whereby farmers could top up their basic area payment. This marked an important change of emphasis in the way the government administered the LFA in England - moving away from supporting high cost production towards wider environmental and social objectives.

In practice it is unclear how much environmental benefit the HFA delivered, with some farmers not even realizing that they were undertaking environmental enhancements (ADAS, 2003: 18) and there remains ambiguity over whether the key purpose of the scheme was socio-economic or environmental. While the main justification for support seemed to be the preservation of the upland environment, the HFA was weakly linked to the purchase of specific environmental outputs and included no clear definition of the sustainable farming systems it was trying to support.

The purpose of LFA support and its role in securing sustainable land management was explored further through the implementation of the SP scheme in England and as part of the negotiations on the new Rural Development Programme for England in 2006. The Secretary of State for Environment, Food and Rural Affairs, Margaret Beckett, made a commitment 'to ensure that farms in the uplands have a full opportunity to be rewarded for improved environmental management of their land' (Defra, 2004b) and it was later made clear that government wanted to use the replacement for the HFA as a tool for the delivery of wider public benefits and in particular to contribute to Defra's target to 'care for our natural heritage, make the countryside attractive and enjoyable for all and preserve biodiversity' (Defra, 2006b: 6). Defra subsequently announced that they intended to fully integrate future upland support into Environmental Stewardship and that they were minded to introduce an Upland Entry Level Scheme from 2010 (Defra, 2006c). It was left unclear how this would be implemented in practice and to what extent it would secure additional environmental benefits or merely continue to compensate for agricultural disadvantage. Although there was a commitment to moving away from the compensatory nature of the HFA, it was also stated that the new scheme 'would offer a higher payment than ELS to reflect the higher costs of farming in these areas' (Defra, 2006c).

In light of the requirement on the European Commission to put forward proposals for future LFA approaches (in response to criticisms made by the European Court of Auditors) the UK may have to make fundamental decisions about its use of this mechanism in future. In contrast to support for organic systems, which is now explicitly justified on the basis of the additional environmental public goods these provide, the purpose of LFA policy remains pitched rather uneasily between compensation for natural handicap, support directed at social objectives and an agri-environment payment.

Prospects for more environmentally sustainable agriculture in England

In the decades up until EU accession in 1973, the dominant UK agricultural policy paradigm was one of support and protection for agricultural production. Although this persisted after accession, in the 1980s and 1990s this policy was accompanied by a greater emphasis on environmental protection, often through the mechanism of payments for reducing the intensity of production (Dobbs and Pretty, 2005). Most recently this has developed into a full-scale alternative model for agricultural intervention, shifting away from agricultural production and income support towards a mix of support for the delivery of environmental public goods, and regulation to reduce the impact of negative environmental externalities, as part of a wider move towards a more market-driven agriculture (HM Treasury and Defra, 2005). In contrast, at the same time a model of 'multifunctional' agriculture for Europe has developed, with support from elements in the European Commission, promoting joint social, production and environmental delivery in rural areas (Lowe *et al.*, 2002).

Although in principle there is no inherent contradiction between the use of market-based tools and the pursuit of higher environmental standards, this review suggests that there is a tension between stated domestic government policies for more environmentally sustainable agriculture and wider government policies favouring liberalization and EU reform. Considerable effort has been expended in England on constructing in Environmental Stewardship an agri-environment scheme that has the potential to make a significant contribution to the government's key biodiversity and resource protection priorities. However, securing EU resources to underpin a policy shift in this direction appears to have been a lesser priority in the UK negotiating position at the EU level than securing trade liberalization and maintaining the UK budgetary position. Thus in 2002 the UK agreed to a Franco-German pact, protecting core farm Pillar I subsidies as part of an EU deal to decouple farm payments from production; and in 2005 the UK - as EU President - brokered a budget deal that largely protected the UK rebate whilst scaling back European Commission proposals to expand Pillar II.

At the domestic level tensions are also apparent between government aspirations for more environmentally sustainable agriculture and the deregulation agenda. The emerging UK model for agriculture relies on application of the 'polluter pays provider gets' principle to determine where the line between incentive and regulation is drawn. However, there is some evidence that in practice the line may be more of a gap, with a lack of resources to incentivize improvements in environmental delivery and a reluctance to regulate to secure change. This tension has manifested itself in the development of cross compliance standards attached to the new decoupled farm payments: the principle of requiring a protective buffer alongside hedgerows was accepted as part of the EU requirement for farm payments to be linked to the protection of landscape features but the government accepted the case for exempting hedges in small fields on the basis of the regulatory burden it could impose.

The government has made some significant steps towards setting out and implementing policies that should help make progress towards more environmentally sustainable agriculture. However, where there is tension with other policy objectives progress is much more limited. Some key policy areas where this tension is likely to be seen in the current period include the CAP Health Check in 2008, the EU budget review in 2008/9, and in the role of agriculture in addressing climate change.

The CAP Health Check will look *inter alia* at the future of set-aside, simplification of cross compliance, fund-switching from Pillar I to Pillar II and capping of Pillar I payments. With respect to set-aside, the UK will need to balance the liberalization argument with the desire to retain the benefits to biodiversity that set-aside currently offers. Cross compliance has enabled the UK for the first time to make an effective link between farm payments and baseline environmental standards. However, precedent suggests that the government will have problems resisting the likely industry pressure to weaken these. On the funding issue the UK government has set out clearly its desire to end Pillar I support payments by 2020. It will be interesting to see how this priority plays against the issue of payment capping (to which the UK has signalled clear opposition) and maintaining the overall UK-EU budget position.

One rapidly emerging issue is that of climate change, which presents significant challenges for agricultural policy, requiring responses ranging from reductions in greenhouse gas emissions, production of renewable energy crops and adaptation both in agricultural production and in allowing adaptive land uses such as flood defences and wildlife corridors. It seems likely that the domestic policy response to this will be at least as problematic as that manifested in the shift towards more sustainable agriculture since 1973. However, the precedent of organic farming, where the government has acted to support a system under which clearly identified public goods are delivered, suggests that support for land management practices that address the challenge of climate change will be forthcoming.

Notes

1. In Member States where transposition of directives is not thorough, cross compliance is likely to have a more significant impact, and this may partly explain the Commission's enthusiasm for its use.
2. The authors would like to acknowledge the contribution of Ian Alexander to this section.
3. The authors would like to acknowledge the contribution of Georgina Dobson to this section.

References

ADAS (2003) *The Mid Term Evaluation of the England Rural Development Programme (Hill Farm Allowance).* Report to Defra by ADAS Consulting Ltd. Defra, London, UK.

Bengtsson, J., Ahnstrom, J. and Weibull, A. (2005) The effects of organic agriculture on biodiversity and abundance: a meta-analysis. *Journal of Applied Ecology* 42(2) 261-269.

Cabinet Office (2002) *Farming and Food: A Sustainable Future.* [Accessed Jan 2007] Available from http://archive.cabinetoffice.gov.uk/farming/index/commissionreport.htm

Commission of the European Communities (CEC) (2004) Communication from the Commission to the Council and the European Parliament: European Action Plan for Organic Food and Farming. COM(2004) 415 final. [Accessed 2007] Available from http://ec.europa.eu/agriculture/qual/organic/plan/comm_en.pdf

Commons Hansard Debates [online] (27 June 2006). [Accessed Jan 2007] Available from http://www.publications.parliament.uk/pa/cm200506/cmhansrd/cm060626/text/60626w1227.htm

Council of the European Union (1999) Council Regulation (EC) No. 1257/1999 of 17th May 1999 on support for rural development from the EAGGF and amending and repealing certain regulations. *Official Journal of the European Union* L 160, 42, 26.6.1999, 80-101.

Council of the European Union (2003) Council Regulation (EC) No. 1782/2003 of 29th September 2003 establishing common rules for direct support schemes under the common agricultural policy and establishing certain support schemes for farmers. *Official Journal of the European Union* L 270, 46, 21.10.2003, 1-69.

Council of the European Union (2005) Council Regulation (EC) No.1698/2005 of 20th September 2005 on support for rural development by the European Agricultural Fund for Rural Development (EAFRD). *Official Journal of the European Union* L 277, 48, 21.10.2005, 1-40.

Council of the European Union (2006) Council decision of 20th February 2006 on Community strategic guidelines for rural development (programming period 2007 to 2013). *Official Journal of the European Union* L 55, 49, 25.2.2006, 20-29.

Defra (2002) *Strategy for Sustainable Farming and Food.* Defra, London, UK.

Defra (2004a) *Secretary of State Margaret Beckett's Statement on CAP Reform: Implementation of the Single Farm Payment Scheme in England.* 12.2.2004. [Accessed Jan 2007] Available from http://www.defra.gov.uk/corporate/ministers/statements/mb040212.htm

Defra (2004b) *Written Ministerial Statement by the Secretary of State on Definition of the English Regions for the Single Payment Scheme.* 22.4.2004. [Accessed Jan 2007] Available from http://www.defra.gov.uk/corporate/ministers/statements/mb040422.htm

Defra (2005a) *Partners for Success: a Farm Regulation and Charging Strategy.* Defra, London, UK.

Defra (2005b) *Set-aside Handbook and Guidance for England.* Defra, London, UK.

Defra (2006a) *Single Payment Scheme Cross Compliance Handbook for England.* Defra, London, UK.

Defra (2006b) *Rural Development Programme for England: 2007-2013. Uplands Reward Structure Consultation Document.* Defra, London, UK.

Defra (2006c) *Written Statement by David Miliband on the Future Uplands Reward Structure - 12 December 2006.* [Accessed Jan 2007] Available from http://www.defra.gov.uk/corporate/ministers/statements/dm061212b.htm

Defra (2007a) *Organic Statistics UK.* [Accessed 2007] Available from http://statistics.defra.gov.uk/esg/statnot/orguk.pdf

Defra (2007b) *Organic Statistics Dataset.* [Accessed 2007] Available from http://statistics.defra.gov.uk/esg/statnot/orgdataset1.xls

Dobbs, T. and Pretty, J. (2005) Agri-environmental stewardship schemes and "multifunctionality". *Review of Agricultural Economics* 26(2) 220-237.

Dwyer, J., Baldock, D. and Einschutz, S. (2000) *Cross Compliance under the Common Agriculture Policy.* A report to the Department of the Environment, Transport and the Regions (DETR) by the Institute of European Environmental Policy (IEEP), London, UK.

EN (2003) £2.5m boost for sustaining wildlife and sheep farmers. *English Nature Press Release* EN/03/14. 27.5.2003. [Accessed Jan 2007] Available from http://www.english-nature.org.uk/news/story.asp?ID=470

Eurostat (2002) *Income of the Agricultural Households Sector: 2001 Report*. Eurostat, Luxembourg.

Firbank, L.G., Smart, S.M., Crabb, J., Critchley, C.N.R., Fowbert, J.W., Fuller, R.J., Gladders, P., Green, D.B., Henderson, I. and Hill, M.O. (2003) Agronomic and ecological costs and benefits of setaside in England. *Agriculture, Ecosystems and the Environment* 95, 73-85.

GFA-RACE/IEEP (2003) *The Potential Environmental Impacts of the CAP Reform Agreement*. Report to Defra. Defra, London, UK.

HM Government (2005) *Securing the Future: Delivering UK Sustainable Development Strategy*. Cm. 6467. HMSO, London, UK.

HM Treasury (HMT) and Defra (2005) *A Vision for the Common Agricultural Policy*. HMT and Defra, London, UK.

Hole, D.G., Perkins, A.J., Wilson, J.D., Alexander, I.H., Grice, P.V. and Evans, A.D. (2005) Does organic farming benefit biodiversity? *Biological Conservation* 122, 113-130.

JNCC (2002) *Environmental Effects of the Common Agricultural Policy and Possible Mitigation Measures*. Report prepared for Defra by the Joint Nature Conservation Committee on behalf of the GB statutory conservation, countryside and environment agencies, London, UK.

Lowe, P., Buller, H. and Ward, N. (2002) Setting the next agenda? British and French approaches to the second pillar of the Common Agriculture Policy. *Journal of Rural Studies* 18, 1-17.

NFU (2006) NFU dismayed at RPA incompetence. *NFU Press Release*. 16.3.2006 [Accessed Jan 2007] Available from http://www.nfuonline.com/x5228.xml

Pywell, R. F., Warman, E. A., Hulmes, L., Hulmes, S., Nuttall, P., Sparks, T. H., Critchley, C. N. R. and Sherwood, A. (2006) Effectiveness of new agri-environment schemes in providing foraging resources for bumblebees in intensively farmed landscapes. *Biological Conservation* 129, 192-206.

RSPB (2005a) *The RSPB Annual Review 2004-2005*. RSPB, Sandy, Bedfordshire, UK.

RSPB (2005b) Europe's wildlife lies bleeding after budget battle. *RSPB Media Release*. 19.12.2005.

Soil Association (2005) *Organic Market Report 2005 - Executive Summary* [Accessed 2007] Available from http://www.soilassociation.org/web/sa/saweb.nsf/librarytitles/20686.HTMl/$file/Organic%20market%20report%20executive%20summary%202005.pdf

SEERAD (2004) *Single Farm Payment Scheme Information Leaflet 7*. [Accessed Jan 2007] Available from http://www.scotland.gov.uk/Publications/2004/10/20108/45384

TFA (2004) 2 metre margins: TFA presses for small field exemption. *Tenant Farmers Association Press Release*. 1.10.04. [Accessed 2007] Available from http://www.tfa.org.uk/news.aspx

Chapter 21

Multifunctional agriculture and integrated rural development in Germany: the case of the Regional Action programme

K. Knickel and S. Peter

Introduction

Finding a new balance between societal demands for high environmental quality and the economic pressures that are resulting from competition in world markets is a key issue in current debates surrounding the idea of 'sustainable agriculture'. New development models, such as the European Landscape Convention (2000), aim to reconcile agricultural production with the maintenance of biological and landscape diversity (Ministère de l'Ecologie, 2004: 1). This Convention acknowledges, for instance, that agriculture can provide environmental amenity and contribute to the maintenance of the cultural heritage, whilst fostering the economic viability of rural areas (Knickel *et al.*, 2004a). Agricultural production systems in Europe range from intensive and often specialized systems, which are competitive on international markets while taking only basic responsibility for biological and landscape diversity, to agricultural systems in more marginal areas that contribute to maintaining and developing a rich diversity of nature and landscape. Pretty (2002) and Hoffmann (2000) argue that agriculture contributes to landscape and nature preservation, not in spite of but *through* land use. 'High nature value farmland' has become an important policy focus, as measures such as the *Pan-European Biological and Landscape Diversity Strategy* (PEBLDS) (1995) (Council of Europe *et al.*, 1996) or the Biodiversity Action Plan for Agriculture (CEC, 2001), amongst others, testify. The sixth EU Environmental Action Programme (2001-2010) aims to halt biodiversity decline by 2010 by conserving high nature value farmland, a descriptor which covers between 15 and 25% of agricultural areas in Europe (European Parliament and Council of the EU, 2002). Meanwhile, the Common Agricultural Policy (CAP) has become the key policy framework for the conservation of this category of farmland, as it obliges member states to implement agri-environmental schemes through its 'second pillar' as well as making support increasingly conditional upon environmentally sound management (European Environment Agency (EEA), 2004). Since the 1992 MacSharry reforms of the CAP farmers who, on a voluntary basis, provide environmental services to protect and enhance the quality of the natural environment, including biodiversity, have been financially rewarded. The regulatory environment thus acknowledges that farmers have an important function as stewards of the environment and the countryside.

In this chapter we discuss the interrelations between the multifunctionality of agriculture and regional development, with a specific focus on Germany and its Regional Action programme. The model regions involved in this pilot programme suggest that innovative integrated forms of rural development are possible. The Regional Action programme is one of three major initiatives that constitute the core of the German Sustainable Development Strategy. The approach tested in the programme secures the two central principles of sustainable regional development, stimulating innovation in a goal-oriented manner and allowing for subsidiarity (WWF, 2002: 65). New societal demands expressed, for example, in the Biodiversity Convention (1992), the EU Flora-Fauna-Habitat Directive (92/43/EEC) and the EU Water Framework Directive (2000) are being addressed. In many projects the guiding principle is the decoupling of economic growth from increased resource consumption and, further still, the active creation of synergies between environmental interests and economic development.

Multifunctionality as a key feature of the 'European Model of Agriculture'

With the Amsterdam Treaty of 1999, the Agenda 2000 and the European Council of Luxemburg in 2003, the European Union made issues of sustainability and multifunctionality key objectives of the CAP (Knickel *et al.*, 2004a: 97). Agriculture and rural areas are perceived not only as producers of commodities but also of non-commodity environmental and social goods. The objectives of the so-called 'European Model of Agriculture' have been described by the European Commission (2003) as:

- A modern and competitive farming sector, capable of occupying a leading position in the world market, while safeguarding domestic producers' living standards and incomes.
- A sustainable, efficient farming sector that uses hygienic, environment-friendly production methods and gives consumers the quality products they desire.
- A farming sector that serves rural communities, reflecting their rich tradition and diversity, and whose role is not only to produce food but also to guarantee the viability of the countryside as a place to live and work, and as an environment in itself.
- A simplified agricultural policy, where the lines are clearly drawn between what is decided at Community level and what is the responsibility of the Member States.

Within this new paradigm, production remains only one among various desired functions of agriculture (Pretty, 1998; Knickel, 2001). In these terms, the role of agricultural land is being redefined and farmers will need to acquire new skills in order to fulfil what Hervieu (2003: 4) calls a 'synthesis profession'. The guiding idea is that through multifunctionality agricultural enterprises can create a broader basis of income generation and at the same time gain greater appreciation of their outputs from society (Knickel, 2001; Künast, 2001; Wilson, 2001). Besides food and non-food crops agriculture produces environmental, social and cultural non-commodity outputs (Meister, 2001; Hervieu, 2003). As the market does not reward most of the latter financially, it is argued that policy should step in to fill this gap (Hervieu, 2003). Open space amenities can generally be seen as non-commodity outputs of agriculture (Abler, 2001; Winter, 2001). At the same time, however, the aesthetic and cultural value of agriculturally shaped landscapes can to a considerable extent be turned into monetary value, for example, by realizing the potential of rural tourism (Knickel, 2001; Pretty, 2002; Knickel and Peter, 2005). Thus, non-marketable agricultural outputs may contribute indirectly to economic growth by helping to enhance the image of a regional landscape.

Several authors observe that, contrary to the tendencies of rural depopulation and economic decline, there are rural areas benefiting from growth in population, employment and income. Much of this growth is stated to be a result of the use of amenity resources (Hunter *et al.*, 2004). In particular, natural resources in the form of landscape elements and formations suitable for recreation are believed to have a significant economic impact (Henderson and McDaniel, 1998; Deller *et al.*, 2001; Green, 2001; Green *et al.*, 2005).

The example of the Regional Action pilot programme

Mobilizing specific regional potentials for developing new economic activities

Against the background of these trends and emphases in agricultural policy, significant regional differences between rural areas need to be taken into account. Rural areas are characterized by a wide diversity of natural, cultural and structural conditions at the regional level; an issue that has rarely been given sufficient consideration by policymakers in the past, but one that the policies of local development agencies, national governments and the EU must reflect when promoting new economic activities (Marsden, 2003). It is in this vein that new bottom-up policy approaches in support of sustainable development of rural areas are being tested.

The German 'Regional Action - Rural Areas Shaping the Future' pilot programme is an example of a new type of support scheme that specifically addresses the development of economic activities as well as their linkages with the enhancement of environmental quality. It was initiated by the German Federal Ministry of Consumer Protection, Food and Agriculture (BMVEL)[1] in 2001. The programme is expected to provide a concrete translation of the concept of the so-called *'Agrarwende'* (agricultural turnaround) into practice (Knickel, 2006).[2] The *Agrarwende* was announced in the course of the reorientation of German agricultural policy as a response to massive pressures resulting from the Bovine Spongiform Encephalopathy (BSE) crisis. Correspondingly, one of the main differences to the EU Leader programme initiated in 1991 (EU, 2004) is Regional Action's explicit focus on the reorientation of agriculture towards sustainability and quality production, nature conservation and rural amenities. Particularly interesting in this respect is its aim of adding value to natural resources while using them more efficiently and at the same time sustainably. Furthermore, the improvement of producer–consumer relations is to be achieved by creating transparent methods of production and marketing; counteracting the growing alienation of consumers from food production (Hervieu, 2003: 1) and the reduction of farming to merely one link in an increasingly industrialized food supply chain. Transparent high quality food production is now understood to be a central starting point for a renewed 'social contract' because it can secure the economic basis of agriculture whilst increasing social investments and appreciation in it (Nachhaltigkeitsrat, 2001).

Beyond promoting the agricultural sector, Regional Action aims to foster the integrated and sustainable development of rural areas. One of its key ideas therefore is to better co-ordinate and strengthen the various functions of rural areas. The active generation of these synergies is central to its activities and their combination at farm and regional level. While specialization in agricultural production and segregation of agriculture from other rural activities had been encouraged in the past, multifunctional and amenity-led development is focused on mutual benefits and 'win–win situations' created by a range of different activities (Brunori and Rossi, 2000; Knickel and Renting, 2000). One of the explicit objectives of Regional Action, for instance, is the expansion of 'tourism services to provide opportunities for combining environmentally compatible agriculture and forestry in an intact landscape with marketing of regional products', thus addressing the interconnectedness of positive landscape image, regional identity and the success of new services and products (BMVEL, 2002: 3). The revitalization and strengthening of urban-rural linkages, such as through the promotion of recreation among urban populations, is a closely related aim, promising mutual benefits.

Integrated regional development through 'regional partnerships'

Regional Action provides support for the realization of micro-region-specific integrated development concepts, jointly devised by regional level actors, institutions and stakeholders (BMVEL, 2002: 7). The size of the 18 model regions ranges from 320 km^2 to 5,800 km^2, the average being 1,500 to 2,000 km^2. The regions have between 36,000 and 1.2 million inhabitants. 'Regions' within the pilot programme's framework are defined as forming functionally and/or spatially homogenous areas - often coherent landscape units - with common problems and potentials, and may encompass several municipalities and administrative districts (BMVEL, 2001a: 1). There are examples of geological boundaries such as mountain ranges or rivers shaping a regional identity (BMVEL, 2002: 13) in spite of the structural heterogeneity of the administrative districts included. Starting conditions varied among the model regions. While several already had a clear regional identity before the implementation of Regional Action, others had yet to develop and communicate a regional identity and build up cooperation structures (Knickel *et al.*, 2004b: 9f).

The model regions were selected competitively. Over 200 applications were made and 18 were chosen on the basis of the quality of their development concepts at the beginning of 2002. The design of the pilot programme as a competition aimed at encouraging greater innovativeness of development concepts and methods of implementation. The winning regions received an annual grant of approximately 1.5 million euros over the period from 2002 to 2007. The Federal Ministry provided a support framework actively backing up regional development activities, including a regional management structure and the implementation of innovative core projects (BMVEL, 2002: 9). Detailed steering was delegated to the regional level, in accordance with the principle of subsidiarity. The experience gained indicates that this approach serves the effective realization of regional environmental and social objectives that tend to be neglected at the state level, though overall federal government interests are accommodated (Fürst, 2001a; 2001b).

Co-operation structures - the so-called 'regional partnerships', representative of those who are actively involved and holding decision-making power - had to be conceptualized as a part of the regional development concepts. Subsequently they provided the organizational basis for the implementation of the programme (BMVEL, 2002: 9). A regional public body, often the district authorities or the agricultural office, was responsible for budget administration, while regional management teams played a key role as agents of networking and skill building (Knickel *et al.*, 2004b: 13ff). As Champetier and Janot (1997) state, partnerships can function as better motors of innovation than single institutions because of their higher potential to integrate different perspectives and competences. Thus, they are becoming increasingly important as the basis of regional development processes (Enright, 1996; Olukosi, 1996; Böcher, 2002).

The role of environmental amenities within the Regional Action programme

Managing the countryside as a cross-sectoral task

Managing the European countryside is less and less understood to be the responsibility of farmers alone. It is a cross-sectoral task involving various other types of actors as well, not least nature conservation and community based organizations. Indeed, Green *et al.* (2005) underline that a community-oriented planning approach is needed in order to allow for the linkage of amenities and rural economic development - a precondition fulfilled by the pilot programme's approach.

Conservation practices enhancing the valorization of environmental assets do not only contribute to the economic competitiveness of agricultural enterprises. The integrated approach of the pilot programme also aims at reconciling diverse regional interests. The idea is that natural amenities benefit because actors from different sectors, such as nature protection, agriculture and tourism, co-operate in joint projects (see English *et al.*, 2000 on the role of amenities for tourism). Thus, natural amenities contribute further added value through environment-friendly use while their viability is secured. However it is as yet unclear to what extent the valorization of natural amenities through their maintenance, enhancement and use can carry regional economic development.

Rayment and Dickie (2001) underline the significant positive impacts of nature conservation on employment and local economies in the UK. Besides direct employment in the agricultural sector local rural tourism benefits from an intact environment. A more thorough analysis of these interrelations has been carried out for the Rhön (Knickel, 2001). The developments in the Rhön are characterized by the active construction of synergies at the farm household, farm and regional level. The maintenance and creation of ecological values (at landscape level) contributes to individual activities in the field of recreation and tourism in the sense of a positive feedback - that is, the overall potential is continuously increased. The development of synergies at the farm household level is expressed in the joint establishment of regional labels and quality products, the processing of own produce, direct marketing and their combination with farm holidays. The enhanced potential for green tourism is expressed in the fact that restaurants and tourist associations in the Rhön are advertising with the positive image and label of the Biosphere Reserve Rhön. Tourists spend money in hotels, restaurants, sightseeing places and so forth, resulting in additional regional income and employment. Total turnover in tourism in the region is approximately €160 million. If it is assumed that at least one third of all tourists come to the area mainly because of its pronounced rural (green) image, then this corresponds to a very considerable economic impact.

Project strategies developed in the model regions

The conservation of rural landscapes that are attractive for living and tourism is an important goal in all of the model regions. The landscape potential of these regions in terms of lakes and flower-rich grasslands and infrastructures, such as the provision of hiking trails, offer visitors and the local population possibilities for leisure activities. These interrelations are reflected in the main aims of the projects implemented (Table 21.1). Of almost 730 projects that have been implemented in the 18 model regions since 2002, the development of eco-tourism plays a major role for over a third (37%), closely followed by agriculture and nature conservation initiatives (34%). Two other important project concerns are information and qualifications (50%) and regional and direct marketing (42 %). In accordance with the integrated approach of the pilot programme, many projects have more than one major concern, actively linking different development strategies with a range of beneficiaries (Table 21.1).

Table 21.1: Contents and beneficiaries of Regional Action projects

Contents and beneficiaries of Regional Action projects	Number of projects	Share of projects (%)
Project can be categorized as...		
...concept development, planning, information, facilitation	523	72%
...investment	137	19%
...regional agri-environmental measures	30	4%
Other	36	5%
Total	**726**	**100%**
Main concerns of the project are...		
...agriculture & nature conservation	247	34%
...regional and direct marketing	305	42%
...non-food production and renewable sources of energy	161	22%
...eco-tourism	271	37%
...information, qualification	360	50%
Other	34	5%
Total	-[a]	-[a]
Direct beneficiaries of the project are...		
...communal bodies	515	71%
...farmers	410	56%
...small and medium enterprises	274	38%
Other	178	25%
Total	-[a]	-[a]

Source: Authors' compilation based on BMVEL, 2001b, 5 December 2005
[a] Projects could be counted under more than one category

Table 21.2: Role of landscape/environment in Regional Action projects

Project aims at .... [a]	**Number of projects**	**Share of projects (%)**
... maintaining/enhancing biodiversity	27	4%
... maintaining/enhancing landscape character/diversity	89	12%
... maintaining/enhancing cultural heritage of landscape/region	33	4%
...improved interrelations between nature and agriculture	119	16%
...improved interrelations between nature and regional development	33	5%
...adding value to natural resources	95	13%
... adding value to landscape (generally)	12	2%
...adding value to landscape through eco-tourism development	117	16%
Other	201	27%
Total	**726**	**100%**

Source: Own compilation based on BMVEL 2001b, 05 December 2005
[a] Projects could be counted under more than one category

In 73% of all projects, adding value to landscape and other natural resources as well as their enhancement or maintenance are clearly recognizable, explicit objectives (Table 21.2). A share of 13% of all projects is mainly aimed at adding value to natural resources, such as wild plants (berries, herbs, etc.) or biomass. Adding value to regional landscapes is a major aim for 18% of all projects, mostly through eco-tourism (16%). Maintaining or enhancing the character and diversity of regional landscapes plays a major role for 12% of projects, while a small share *primarily* focus on the cultural heritage of regional landscapes (5%) or maintaining or increasing biodiversity (4%).

Due to this integrated, intersectoral approach, the projects are addressing development processes that take place at several interfaces. A share of 16% of all projects is located at the interface between agriculture and landscape/nature, while 5% are sited at the interface between nature and the overall (economic) development of the region (Table 21.2).

A further set of projects address environmental management goals. The regional variations of agriculture and of types and intensities of land use demand a regionally differentiated agri-environmental support framework reflecting the diversity of the environment in order to meet nature conservation objectives. An important question therefore is how regional and local level actors can be involved in the differentiation and fine-tuning of national schemes without leading to a disproportionate increase in administration and control costs. Some projects implemented in the model regions aim at the development and implementation of region-specific landscape and biotope management measures that simultaneously contribute to the maintenance of regional cultural landscapes (including high nature value grassland) while at the same time creating new income opportunities for farm households. A closely related strategy is the establishment of regional marketing initiatives and regional umbrella brand names under which products from extensive agriculture are marketed. While such projects account for only 4% of all projects implemented so far, they are nonetheless important steps in terms of a valorization of the environmental qualities of the region.

Many of the projects implemented in the model regions innovatively address the objectives of creating new sources of income for rural areas through diversification and increasing regional value added while preserving and valorizing natural resources or amenities. The strong cross-sectoral alignment of such projects mirrors the integrated overall approach of the pilot programme. A frequent constellation is alliances connecting landscape, tourism and gastronomy.

Distinctive nature and landscape potentials encompassing rare species such as the stork, ice-age landscape elements of outstanding geo-historical meaning, or typical flower meadows are being re-assessed with regard to their economic, ecological and aesthetic value, allowing for different interests. Nature-friendly farm-based tourism is an example. Most farm households involved in tourism are offering their own quality products to their guests. For many tourists, this is a major reason why they choose farm holidays.

Natural resources also play an increasing role in the field of renewable decentralized forms of energy generation. Several projects in the model regions aim at the establishment of regional-level networks that can strengthen cross-sectoral cooperation and knowledge transfer regarding the use of renewable sources of energy.

Conclusions

Against the background of a re-orientation of German agricultural and consumer policy, as well as corresponding developments at EU level, the region has become an important unit for the implementation of policy measures as well as pursuing 'new ways' of thinking about rural development (Knickel, *et al.*, 2004a). Both can be detailed further in terms of the complexity of an integrated, economically, ecologically and socially sustainable development which calls for cross-sectoral approaches. As this complexity is still transparent at the regional level and because of the proximity to practice and its problems, as well as the direct perception of interrelations, changes and impacts, the recognition of synergies is likely to be achieved more effectively from within the region (Brunori and Rossi, 2000; WWF, 2002; Baier and Bennholdt-Thomsen, 2003).

From practice to theory

Model regions and model projects show what abstract concepts such as sustainable development or the concept of multifunctional agriculture and rural space could mean in actual practice (van der Ploeg *et al.*, 2002; Knickel *et al.*, 2004a; Peter and Knickel, 2006). At the same time careful and open-minded analysis of practical experiences can enrich the development of theory. The integrated development approach pursued by the Regional Action programme contributes to intersectoral cooperation by facilitating the creation of new alliances between various groups and joint action. Newly formed cooperations have resulted in a considerable number of projects emphasizing the valorization and, at the same time, maintenance of natural resources and amenities respectively. Enhancing regional landscapes is as much part of this as the use of renewable energy sources.

The characterization of the many new fields of activities that contribute to multidimensional developments in the model regions is in line with the view that a strict segregation of different functions (for example, living, production, recreation and nature conservation) is less and less realizable. The new forms of multifunctionality, taken together, can result in the construction of a new resource base at the regional level (Knickel and Renting, 2000; Knickel, 2001). An important question in this context, raised by Green *et al.* (2005) is how resource management, tourism and economic development planners could better integrate their strategies in order to enhance the benefits of their cooperation and at the same time counteract interest conflicts and negative externalities.

Regional identity shaped by natural assets as a rural development factor

As described above, the definition of their territorial boundaries was an important precondition for the regions' participation in the competitive scheme (BMVEL, 2001a: 1). The particular challenge lay in the fact that regions as geographical areas do not necessarily equal areas of administration or a space with a shared identity concept (Knickel *et al.*, 2004b: 9).

Regional identity connected to the specific natural environment plays a key role for the developments initiated by the pilot programme. Regional products are advertised with a high nature quality image which is in line with the promotion strategy of tourism agencies and the local catering and restaurant sector. This image supplies a concept creating coherence between the various activities and needs to be communicated to potential consumers in order to increase regional value added through region-specific services and products, for example, under a regional umbrella brand name (Künast, 2001: 6).

Local and regional level actors see the greatest threat to these new developments in the dominant trend towards a concentration and globalization of the larger food economy. These processes clearly put substantial countervailing pressure on quality, price and cultural and regional distinctiveness. Regional shops that provide the population with high quality foodstuffs of local origin have to compete with an extremely centralized retail sector, with mega-retailers and food service companies acting as effective gatekeepers to the entire agri-food chain (EEA, 1998; Marsden, 2003; Kirwan *et al.*, 2004).

Towards a reconstitution of nature-society relations

There have been a number of developments in society leading to a re-evaluation of agricultural and rural development goals and prompting initiatives such as the Regional Action programme. First of all, food production has become a less important issue in quantitative terms. Repeated crises related to food quality clearly demonstrate that today agricultural production is being assessed more in terms of food quality and safety than quantity (Von Alvensleben, 2000). Second, environmental questions, the 'consumption' of nature and standards of living in rural areas have become much more important (Knickel and Peter, 2005). The idea of sustainability has led to a reassessment of the use of natural resources and, as a result, less intensive and more diversified forms of agriculture are now well regarded because they tend to be better adapted to natural conditions and because of the more favourable conditions they provide for an integrated development of rural areas (Knickel, 2001).

The projects that are being implemented as well as the entire Regional Action initiative indicate that agriculture and, more generally, the potential of rural areas are no longer being evaluated in monofunctional terms. There is also sufficient consensus that, without farming, the maintenance of high nature value areas and semi-natural pastoral habitats would hardly be feasible or would be very costly. The synergies between farming and nature conservation become very strong when both sides benefit from it (Knickel, 2001). In this respect the rediscovery of the multifunctionality of agriculture is mainly a result of societal changes which have been taken up by policymakers. There has been a rediscovery of farming as more than just a monofunctional activity involving (food and non-food) production and that - more than other economic activities - it can produce a range of goods and services including those amenities that are appreciated by society but that do not have a real price in the market. Regional actors perceive a balanced economic development as a precondition of strengthening the role of farmers as producers of services, landscapes and biodiversity.

Notes

1. Since November 2005 it is called Federal Ministry of Food, Agriculture and Consumer Protection (BMELV).
2. The Institute for Rural Development Research (IfLS) at Goethe University Frankfurt carried out the accompanying research for policymakers on the pilot programme during the years 2002 and 2003.

References

Abler, D. (2001) Multifunctionality: the question of jointness, applying the OECD framework: a review of literature in the United States. Background paper prepared for the OECD conference *Multifunctionality: Applying the OECD Analytical Framework Guiding Policy Design*, 2-3 July, Paris, France. [Accessed Feb 2005] Available from http://www.oecd.org/agr/mf/doc/USA-Abler-revised.pdf

Baier, A. and Bennholdt-Thomsen, V. (2003) Der 'Stoff', aus dem die soziale Nähe ist. In: Kluge, T. and Schramm, E. (eds) *Aktivierung durch Nähe, Regionalisierung nachhaltigen Wirtschaftens.* Ökom-Verlag, München, Germany, 12-21.

BMVEL (2001a) *Von der Vision zur Realisierung innerhalb einer neu ausgerichteten Agrarpolitik Federal.* Ministry of Consumer Protection, Food and Agriculture, Bonn, Germany.

BMVEL (2001b) Projektdokumentation. [Accessed Dec 2005] Available from http://www.modellregionen.de/home.htm

BMVEL (2002) *Active Regions - Shaping Rural Futures, Competition Winners.* Federal Ministry of Consumer Protection, Food and Agriculture, Bonn, Germany.

Böcher, M. (2002) Nachhaltige Regionalentwicklung durch Kooperation? Erfolgsfaktoren für regionale Partnerschaften aus politikwissenschaftlicher Sicht. In: Müller, K., Dosch, A., Mohrbach, E., Aenis, T., Baraneck, E., Boeckmann, T., Siebert, R. and Toussaint, V. (eds) *Wissenschaft und Praxis der Landschaftsnutzung, Formen Interner und Externer Forschungskooperation.* Margraf Verlag, Weikersheim, Germany, 65-75.

Brunori, G. and Rossi, A. (2000) Synergy and coherence through collective action, some insights from Tuscany. *Sociologia Ruralis* 40(4) 409-423.

Champetier, Y. and Janot, J-L. (1997) *Aufbau und Organisation lokaler Partnerschaften.* Series *Innovation im ländlichen Raum 2.* Europäische Beobachtungsstelle LEADER (AEIDL), Brussels, Belgium.

Commission of the European Communities (CEC) (2001) Communication from the Commission to the Council and the European Parliament: Biodiversity Action Plan for Agriculture. COM(2001) 162 final. [Accessed 2007] Available from http://eur-lex.europa.eu/LexUriServ/LexUriServ.do?uri=CELEX:52001DC0162(03):EN:HTML

Council of Europe, UNEP and European Centre for Nature Conservation (1996) *The Pan-European Biological and Landscape Diversity Strategy*. Council of Europe, Strasbourg, France.

Deller, S.C., Tsung-Hsiu, T., Marcouiller, D.W. and English, D.B.K. (2001) The role of amenities and quality of life in rural economic growth. *American Journal of Agricultural Economics* 83, 352-365.

English, D.B.K, Marcouiller, D.W. and Cordell, H.K. (2000) Linking local amenities with rural tourism incidence, estimates and effects. *Society and Natural Resources* 13, 185-202.

Enright, M.J. (1996) Regional clusters and economic development, a research agenda. In: Staber, U.H., Schaefer, N.V. and Sharma, B. (eds) *Business Networks: Prospects for Regional Development*. Walter de Gruyter, Berlin, Germany and New York, USA, 190-213.

EU (2004) Europa - gateway to the European Union. [Accessed Jan 2005] Available from http://europa.eu.int/comm/agriculture/rur/leaderplus/index_en.htm

European Commission (2003) What is the European model of agriculture? COM, Brussels, Belgium. [Accessed Mar 2003] Available from http://europa.eu.int/comm/agriculture/faq/q1/index_en.htm

European Environment Agency (1998) *Europe's Environment: the Second Assessment*. Elsevier Science Ltd., Oxford, UK.

European Environment Agency (2004) High Nature Value Farmland: Characteristics, Trends and Policy Challenges. [Accessed Jan 2005] Available from http://reports.eea.eu.int/report_2004_1

European Parliament and Council of the European Union (2002) Decision No.1600/2002/EC of the European Parliament and of the Council of 22nd July 2002 laying down the Sixth Community Environment Action Programme. *Official Journal of the European Union* L 242, 45, 10.9.2002, 1-16.

Fürst, D. (2001a) *Regional Governance Zwischen Wohlfahrtsstaat und Neoliberaler Marktwirtschaft*. [Accessed Jan 2005] Available from http://www.laum.uni-hannover.de/ilr

Fürst, D. (2001b) Regional governance, ein neues Paradigma der Regionalwissenschaften. *Raumforschung und Raumordnung* 59(5/6) 370-380.

Green, G.P., Marcouiller, D.W. and Deller, S. (eds) (2005) *Amenities and Rural Development: Theory, Methods and Public Policy*. Edward Elgar Publishing, Northampton, Massachusetts, USA.

Green, G.P. (2001) Amenities and community economic development: strategies for sustainability. *The Journal of Regional Analysis and Policy* 31, 61-75.

Henderson, J. and McDaniel, K. (1998) Do scenic amenities foster economic growth in rural areas? *Regional Economic Digest* 1st quarter, 11-17.

Hervieu, B. (2003) *Multifunctionality: a Conceptual Framework for a New Organization of Research and Development on Grasslands and Livestock Systems*. Institut National de la Recherche Agronomique (INRA), Paris, France.

Hoffmann, L.B. (ed.) (2000) *Stimulating Positive Linkages between Agriculture and Biodiversity. European Centre for Nature Conservation* (ECNC), Tilburg, The Netherlands.

Hunter, L.M., Boardman J.D. and Saint Onge, J. (2004) *The Association between Natural Amenities, Rural Population Growth and Long-term Residents' Economic Well-being*. Research Program on Environment and Behavior, Working Paper EB2004-0005. [Accessed Nov 2005] Available from http://www.colorado.edu/ibs/pubs/eb/eb2004-0005.pdf

Kirwan, J., Slee, B. and Foster, C. (2004) *A Descriptive and Analytical Macro-level Overview of the Dynamics and Diversity of Food Supply Chains in Europe in Relation to their Institutional Setting*. SUSCHAIN WP2 Synthesis Report. Countryside and Community Research Unit, University of Gloucestershire, Cheltenham, UK.

Knickel, K. (2001) The marketing of Rhöngold milk: an example of the reconfiguration of natural relations with agricultural production and consumption. *Journal of Environmental Policy and Planning* 3, 123-136.

Knickel, K. (2006) Agrarwende, agriculture at a turning point: rural development impact and policy in Germany. In: Gorman, M., Renting, H., Kinsella, J. and O'Connor, D. (eds) *Driving Rural Development in Europe: the Role of Policy*. Van Gorcum, Assen, The Netherlands, 82-110.

Knickel, K. and Peter, S. (2005) Amenity-led development of rural areas: the example of the Regional Action pilot program in Germany. In: Green, G.P., Marcouiller, D.W. and Deller, S. (eds) *Amenities and Rural Development: Theory, Methods and Public Policy.* Edward Elgar Publishing, Northampton, Massachusetts, USA, 302-321.

Knickel, K. and Renting, H. (2000) Methodological and conceptual issues in the study of multifunctionality and rural development. *Sociologia Ruralis* 40, 512-528.

Knickel, K., Renting, H. and van der Ploeg, J.D. (2004a) Multifunctionality in European agriculture. In: Brouwer, F. (ed.) *Sustaining Agriculture and the Rural Economy: Governance, Policy and Multifunctionality.* Edward Elgar Publishing, Cheltenham, UK and Northampton, Massachusetts, USA, 81-103.

Knickel, K., Siebert, R., Ganzert, C., Dosch, A., Peter, S. and Derichs, S. (2004b) *Wissenschaftliche Begleitforschung des Pilotprojektes Regionen Aktiv - Land gestaltet Zukunft, Ergebnisse der Begleitforschung 2002-2003.* IfLS, Frankfurt/M, Germany.

Künast, R. (2001) Grußwort. [Accessed Jan 2005] Available from http://www.modellregionen.de

Marsden, T. (2003) *The Condition of Rural Sustainability.* Van Gorcum Press, Assen, The Netherlands.

Meister, A.D. (2001) Synthesis and evaluation of the evidence from the country case studies concerning different arrangements and institutional options for providing non-commodity outputs. Paper prepared for Workshop on Multifunctionality, Paris, 2-3 July 2001. Directorate for Food, Agriculture and Fisheries, Paris, France.

Ministère de l'Ecologie et du Developpement Durable (2004) From knowledge of landscapes to landscaping action. Announcement paper of the International Conference of the French Ecology and Sustainable Development Ministry, 2-4 Dec. 2004. Bordeaux, France.

Nachhaltigkeitsrat (2001) *Pilotprojekt Multifunktionale Landwirtschaft in der Region Umweltbundesamt (UBA)*, Bonn and Berlin, Germany.

Olukosi, J.O. (1996) Local organizations and innovation. In: Sebillotte, M. (ed.) *Systems-oriented Research in Agriculture and Rural Development: Lectures and Debates.* Proceedings of International Symposium, 21-25 Nov 1994. Montpellier, France, 362-364.

Peter, S. and Knickel, K. (2006) Empowerment of regional partnerships: the example of the Regional Action pilot programme in Germany. *DISP-Online* 166, 42(3) 16-25. [Accessed 2007] Available from http://www.nsl.ethz.ch/index.php/en/content/view/full/1219/

Pretty, J. (1998) *The Living Land: Agriculture, Food and Community Regeneration in Rural Europe.* Earthscan, London, UK.

Pretty, J. (2002) Change in agricultural policy and its consequences: will conservation keep farmers in business? In: Frame, J. (ed.) *Conservation Pays?* BGS Occasional Symposium No.36. British Grassland Society, University of Reading, Reading, UK, 15-25.

Rayment, M. and Dickie, I. (2001) *Conservation Works... for Local Economies in the UK.* The Royal Society for the Protection of Birds (RSPB), Bedfordshire, UK.

van der Ploeg, J.D., Long, A. and Banks, J. (2002) *Living Countryside: Rural Development Processes in Europe, the State of the Art.* Elsevier, Doetinchem, The Netherlands.

Von Alvensleben, R. (2000) Zur Bedeutung von Emotionen bei der Bildung von Präferenzen für regionale Produkte. *Agrarwirtschaft* 12, 399-402.

Wilson, G.A. (2001) From productivism to post-productivism ... and back again? Exploring the (un)changed natural and mental landscapes of European agriculture. *Transactions of the Institute of British Geographers* 26(1) 77-102.

Winter, M. (2001) *Multifunctionality: Applying the OECD Framework. A Review of Literature in the United Kingdom.* OECD, Directorate for Food, Agriculture and Fisheries, Paris, France.

WWF (2002) *Entwicklung einer Methodik zur Nutzung des Instruments Globalzuschuss gemäß EU-Strukturfonds.* Final report 900 19 153/02. Umweltstiftung WWF Deutschland, Berlin, Germany.

Chapter 22

Agendas for transdisciplinarity

R. Fish, S. Seymour, C. Watkins and M. Steven

Like the idea of sustainability, the history of transdisciplinarity has often been written as one of growing significance since the 1960s and 1970s; an idea whose time has come. This is no coincidence, for it is partly in response to the widely felt imperatives of sustainable development that investments in this idea have found expression. Transdisciplinarity is often characterized as a problem-orientated and socially reflexive form of academic endeavour; one driven by the needs of application (Gibbons *et al.,* 1994). It is therefore not surprising that deep seated tensions between economic growth, social reproduction and environmental change have been an important backdrop to the development of transdisciplinary work.

Yet, precisely what this work might entail in terms of a coherent ontology or a recognizable set of research practices for the study of sustainable farmland management seems less clear. Certainly, advocates of transdisciplinarity seek, by definition, to avoid disciplinary reductionism, arguing that compartmentalized forms of knowledge are ill-equipped to deal with a world that is messy, uncertain and complex. Transdisciplinarity, we are told, is about amplifying understandings of the world in ways that could not be anticipated when working in disciplinary isolation; to create no less than a 'science and art of discovering bridges between different areas of knowledge and different beings' (Thompson-Klein, 2004: 515). In making this case, advocates often employ a teleological narrative whereby transdisciplinary research is the logical development on 'weaker' conceptions of disciplinary transgression (Balsinger, 2004). Typically, this is a story in which different communities of scholarship 'arrive' at transdisciplinarity by way of 'multidisciplinarity' (the provision of a series of distinct disciplinary perspectives around a given research problem) and 'interdisciplinarity' (the development of common methodological frames around which disciplinary work interacts and synthesizes its outcomes). Transdisciplinarity is thus positioned at the apex of a process in which disciplinary registers and conventions have begun to dissipate, and from which it may be possible to begin envisaging a 'general theory of systems and structures' (Thompson-Klein, 2004: 515). Indeed, it is in this vein that David Harvey has reasoned in the context of rural economy and land use that, 'interdisciplinary methods might be capable of development to a transdisciplinary state, involving unification of the involved disciplines at the paradigmatic (metaphysical) level' (Harvey, 2006: 332).

It is tempting in the context of this book's problematic to craft this idea of unification in the image of the sustainability agenda itself: the neatly overlapping spheres of economy, society and environment replaced by disciplinary labels and implying some kind of idealized 'middle ground' of academic discourse, with a common set of methods and a shared sense of purpose. But like discourses of sustainable development, who could say what would define such a ground? The diversity of approaches and perspectives employed in this volume imply that it is perhaps too simplistic to attempt straight-jacketing transdisciplinarity into one particular model of working, or one all encompassing ontology and, as Adger *et al.* (2003) suggest, there may be significant drawbacks in attempting to do so. Integration at the paradigmatic level may circumscribe the variety of perspectives from which an issue could be approached, whilst the depth (and inherent incongruities) of disciplinary knowledge may be recognized only superficially. The problem context of this book is telling in this respect. After all, when it is possible to stage an encounter between transdisciplinarity and farmland management through such a wide variety of means - the modelling of 'whole farm' systems; the inspection of emerging networks of agronomic advice; engagement with new food movements; the interrogation of different scales of political reality; the unpacking of moral discourses that surround non-human beings; an assessment of farmers' tacit scientific knowledge; the development of sustainability indicators; the interpretative treatment of long since past attitudes - *when it is possible to do all these things*, it is clear that we must exercise some caution in regarding transdisciplinarity as somehow awaiting final methodological or theoretical definition.

A more modest approach to this idea would thus be to understand transdisciplinarity as a 'point of departure' against which the moral, ecological, technological and political apprehensions of a given problem context may begin to take shape as distinctive, though by no means uniform, programmes of action. Such is the way that Thompson-Klein (2004: 516) writes of transdisciplinarity. It is a process comprising not so much a 'new or super discipline' as an 'open structure of unity' in which ostensibly different levels of reality start to cohere in different ways.

This is indeed a very generous and non-prescriptive interpretation of transdisciplinary though it does beg the question as to the extent to which transdisciplinary approaches to sustainable farmland management might then differ from research of an interdisciplinary or, to some extent multidisciplinary, kind, for is it not precisely this kind of generic appeal that can be used to marshal the case for any range of practices purporting to resist disciplinary reductionism? In this we might usefully introduce a further, recurring, dimension to this trajectory of work, namely that what distinguishes transdisciplinarity from other models of cross-disciplinary endeavour is its tendency to collapse neat distinctions between scholarly and non-scholarly communities of practice. Whereas inter- and multidisciplinary models of working tend to flow 'outwards' from disciplines when devising the nature of a research problem and how knowledge is produced, validated and assigned relevance, transdisciplinarity, it is often claimed, works the other way around. It is governed by an 'outside in' view of intellectual activity, one in which approaches to inquiry are actively shaped by questions of application and an enlargened universe of stakeholder values, politics and practical 'know-how' (Hochtl *et al.*, 2006). Not surprisingly, then, transdisciplinary research has been strongly associated with models of environmental governance and decision making that tend to be participatory and inclusive in style. A characteristic feature of this work is that knowledge tends to be viewed as a process of co-production between the custodians, customers and subjects of research. The work of Lyon and Harris in this volume, where farmers are cast as experts - *scientists* - in their own professional community, is highly indicative of this strand of transdisciplinary research.

In other words, transdisciplinary research comprises not only a process of working across the boundaries of disciplines but across the boundaries of scholarship and practice.

As Nowotny (2006) puts it, 'transdisciplinarity does not respect institutional boundaries'. It is a 'forum or platform' for research that recognizes, celebrates even, the idea that knowledge cannot be 'contained':

> Nobody has anywhere succeeded for very long in containing knowledge. Knowledge seeps through institutions and structures like water through the pores of a membrane. Knowledge seeps in both directions, from science to society as well as from society to science. It seeps through institutions and from academia to and from the outside world (Nowotny, 2006).

In transdisciplinary terms, accounts of sustainable farmland management would therefore make a virtue of straddling and collapsing these boundaries of scholarship and practice.

Liberating and innovative though all of this sounds this is nonetheless a process that carries with it a certain politics. If it is one thing to claim that transdisciplinarity is about creating forms of coherence that are neither reducible to disciplinary structures of knowledge nor to conventional scholarship alone, it also clear that some approaches to research are more amenable to transdisciplinary endeavour than others. Transdisciplinarity, of course, invites a great deal of speculation about what counts as 'the problem', what counts as 'the evidence', what counts as the 'appropriate research method' and what counts as the 'appropriate intervention' or 'solution'. It is therefore ostensibly open to a range of treatments and unanticipated coherences but it is a form of endeavour that, in practice, is inclined to translate the idea of 'real world responsiveness' into a question of instrumental action. This is because, in part, the provenance of transdisciplinarity's 'transgressiveness' is in science, and in particular, in debates concerning the ability of 'normal' science to frame its activities in ways that could act upon the world more directly. In consequence transdisciplinary research has often fashioned itself as a highly systemic and pragmatic response to the idea of 'problem-centred' and 'needs-led' research. And as Horlick-Jones and Sime (2004) explain, this is an idea that therefore carries with it the risk that transdisciplinary models of knowledge production will be limited to largely technocratic views of the world. Balsinger (2004) has written of transdisciplinarity as involving 'mutually enhancing' epistemologies but for Horlick-Jones and Sime (2004) this is a process that can easily sidetrack epistemologies that do not fit neatly into the idea of workable solutions and practical translations. Paradoxically, models of research based on more participatory forms of inquiry (Kindon *et al.*, 2007) - ones that may recognize the subjects of such research as experts in their communities (Irwin, 1995) - can be lost here. In this sense the volume's concern with new transdisciplinarity approaches has been not only about formulating effective, but tightly defined 'process responses', but also about amplifying the endemic uncertainties and trade-offs that come from the idea of 'sustainable farmland management'. It has attended, in a critical fashion, to the different coherences that the border work of transdisciplinarity might afford itself.

References

Adger, W.N., Brown, K, Fairbrass, J., Jordan, A., Paavola, J., Rosendo, S. and Seyfang, G. (2003) Governance for sustainability: towards a 'thick' analysis of environmental decision-making. *Environment and Planning A* 35, 1095-1110.

Balsinger, P.W. (2004) Supradisciplinary research practices: history, objectives and rationale. *Futures* 36, 407-21.

Gibbons, M., Limoges, H., Nowotny, H., Schwartzman, S., Scott, P. and Trow, M. (2004) *The New Production of Knowledge.* Sage, London, UK.

Harvey, D. (2006) RELU special issue: editorial reflections. *Journal of Agricultural Economics* 57, 329-336.

Hochtl, F., Lehringer, S. and Konold, W. (2006) Pure theory or useless tool? Experiences with transdisciplinarity in the Piedmont Alps. *Environmental Science and Policy* 9(4) 322-329.

Horlick-Jones, T. and Sime, J. (2004) Living on the border: knowledge risk and transdisciplinarity. *Futures* 36, 441-456.

Irwin, A. (1995) *Citizen Science: A Study of People, Expertise and Sustainable Development.* Routledge, London, UK.

Kindon, S., Pain, R. and Kesby, M. (2007) *Connecting People, Participation and Place: Participatory Action Research Approaches and Methods*. Routledge, London, UK.

Nowotny, H. (2006) The potential of transdisciplinarity. [Accessed 2007] Available from http://wwww.interdisciplines.org/papers/5

Thompson-Klein, J. (2004) Prospects for transdisciplinarity. *Futures* 36, 515-526.

Index